THE GREEN REVOLUTION

THE GREEN REVOLUTION

The American Environmental Movement,

1962–1992

KIRKPATRICK
SALE

A CRITICAL ISSUE

CONSULTING EDITOR: ERIC FONER

 HILL AND WANG

A division of Farrar, Straus and Giroux / New York

LIBRARY OF CONGRESS CATALOGING-IN-PUBLICATION DATA
Sale, Kirkpatrick.
The green revolution : the environmental movement 1962–1992 /
Kirkpatrick Sale ; consulting editor, Eric Foner.—1st ed.
    p.   cm.—(A critical issue series)
Includes index.
1. Green movement.   I. Foner, Eric.   II. Title.   III. Series.
HC79.E5S253   1993   363.7′057—dc20   92-32607   CIP

To Rachel Carson and those like her
working in the hope that the tragic
prophecy in her dedication of *Silent
Spring* does not come true:
"To Albert Schweitzer,
who said,
'Man has lost the capacity
to foresee and to forestall.
He will end by destroying
the earth.' "

CONTENTS

As for those who would take the whole world
To tinker with as they see fit,
I observe that they will never succeed:
For the world is a sacred vessel
Not made to be altered by man.
The tinkerers will spoil it;
Usurpers will lose it.

Lao-tse

TIMELINE

ORGANIZATIONS		EVENTS		LEGISLATION	
1. Origins					
	1866 American Society for the Prevention of Cruelty to Animals				
	1892 Sierra Club				
	1905 National Audubon Society				
	1919 National Parks and Conservation Association				
	1922 Izaak Walton League				
	1935 Wilderness Society				
	1936 National Wildlife Federation				
	1947 Defenders of Wildlife			1955	Clean Air Act
	1951 Nature Conservancy			1960	Clean Water Act
	1961 World Wildlife Fund			1962	White House Conservation Conference
2. 1962–70		1962	*Silent Spring* (Carson)		
			Our Synthetic Environment (Bookchin)		
			Thalidomide		
		1963	National Park Service "Leopold Report"	1963	Clean Air Act amended
					Partial Nuclear Test Ban Treaty

TIMELINE

ORGANIZATIONS	EVENTS	LEGISLATION
	1964 John Muir stamp issued d. Rachel Carson	1964 Wilderness Act
	1965 Northeast U.S. blackout Storm King case	1965 Water Quality Act Solid Waste Disposal Act Clean Air Act amended
1966 Clearwater project	1966 First Lunar Orbiter photo of Earth NYC air inversion U.S. B-52 with four hydrogen bombs crashed near Spain Unsafe at Any Speed (Nader)	1966 National Wildlife Refuge System
1967 Environmental Defense Fund Fund for Animals	1967 Torrey Canyon oil spill (36 m. gallons), English Channel Three NASA astronauts died in launch-pad fire 100 million vehicles on U.S. roads	1967 Clean Air Act amended

1968 Zero Population Growth	Grand Canyon project ended	1968 National Wild and Scenic Rivers Act
	1968 Paris Biosphere Conference (UNESCO)	National Trails Act
	The Population Bomb (Ehrlich)	Grand Canyon dams bill defeated
	First Club of Rome meeting	1969 National Environmental Policy Act
1969 Council on Economic Priorities	1969 Santa Barbara oil spill (amount undetermined)	
Friends of the Earth	Cuyahoga River bursts into flames	
Union of Concerned Scientists	First SST flight	
	Cyclamates banned, MSG limited	
	So Human an Animal (Dubos)	
	Reith Lectures by F. F. Darling	
	Clearwater launched	
1970 Center for Science in the Public Interest	1970 Nobel Peace Prize to Norman Borlaug for "Green Revolution"	1970 Environmental Protection Agency established
Environmental Action	First complete gene synthesis, U. Wisconsin	Clean Air Act amended
League of Conservation Voters	*Sand County Almanac* (Leopold) reissued	Water Quality Control Act
Natural Resources Defense Council		Occupational Safety and Health Act

TIMELINE

	ORGANIZATIONS		EVENTS		LEGISLATION
	Public Interest Research Groups		Earth Day, April 22		National Industrial Pollution Control Council
1971	Greenpeace	1971	*Calvert Cliffs* case		
	Public Citizen		Cigarette ads banned from TV		
			U.S. exploded hydrogen bomb, Amchitka Island		
1972	Trust for Public Land	1972	UN Stockholm Conference on Human Environment	1972	Water Pollution Control Act
	UN Environment Programme		First returnable bottle law, Oregon		Marine Mammal Protection Act
			Values Party established, New Zealand		Insecticide, Fungicide, Rodenticide (Pesticide Control) Act
			Blueprint for Survival (*Ecologist*)		Coastal Zone Management Act
			Limits to Growth (Meadows et al.)		
1973	Cousteau Society	1973	OPEC oil embargo, "energy crisis"	1973	Endangered Species Act

Year	Organizations	Events	Legislation
1974	Environmental Policy Institute	First "environmental strike," Chemical and Atomic Workers vs. Shell *Small Is Beautiful* (Schumacher)	Trans-Alaska Pipeline Act Safe Drinking Water Act
1974		Karen Silkwood killed, Oklahoma CFC danger reported Cocoyoc Statement (UNEP, UNCTAD) World population at 4 billion *Turtle Island* (Snyder)	
1975	Environmentalists for Full Employment Food First Worldwatch Institute	Salmon return to Connecticut River, sturgeon to Hudson FMC corporation dumped 3,000 pounds of toxic waste into Ohio River	Hells Canyon, Tocks Island bills defeated
1976		Seabrook anti-nuclear demonstrations National Academy of Sciences opposed aerosol sprays Legionnaires' disease outbreak, Philadelphia	Federal Land Policy and Management Act National Forest Management Act Toxic Substances Control Act

TIMELINE

ORGANIZATIONS	EVENTS	LEGISLATION
		Resource Conservation and Recovery Act
		Fishery Conservation Act
1977 Abalone Alliance	1977 Blackout, NYC	1977 Clean Air Act amended
Sea Shepherd	North Sea oil spill (8.2 m. gallons)	Clean Water Act amended
	U.S. admitted neutron bomb testing	Ocean Dumping Act amended
	1978 Love Canal contamination reported	1978 Port and Tanker Safety Act
	Three Mile Island radioactive leak	Energy Policy and Conservation Act
	First test-tube baby, U.K.	
	Amoco Cadiz oil spill off France (70 m. gallons)	
	Politics of Cancer (Epstein)	
	Arrogance of Humanism (Ehrenfeld)	
	Oil drilling begun off New Jersey shore	
1979 Greenpeace	1979 World population at 5 billion	

	Surgeon General Report, *Healthy People*	1980 Superfund (Comprehensive Environment Response Act)
	Oil tanker collision, Trinidad and Tobago (97 m. gallons, largest ever)	Fish and Wildlife Conservation Act
1980 Earth First!	1980 *Global 2000 Report* to President Carter	Alaska National Interest Lands Act
	World Conservation Strategy (IUCN)	Low-Level Radioactive Waste Act
	Overshoot (Catton)	Alternative Motor Fuels Act
1981 Citizens' Clearinghouse for Hazardous Wastes	1981 Reagan Administration installed	1982 Nuclear Waste Policy Act
	EPA settled with Dow on 2,4D case	
1982 Americans for the Environment	1982 EPA scandal	
Bat Conservation International	UN World Charter for Nature	
Co-op America		
Earth Island Institute		
Rocky Mountain Institute		

4. 1981–87

TIMELINE

ORGANIZATIONS	EVENTS	LEGISLATION
1983 Elmwood Institute	1983 Times Beach, Missouri, abandoned	
	Die Grünen elected to parliament, West Germany	
1984 U.S. Greens (Committees of Correspondence)	1984 Bhopal plant disaster, India	1984 Resource Conservation Act amended
National Toxics Campaign	Brundtland Commission appointed	
North American Bioregional Congress I	Famine in the Sahel, Africa	
	First Worldwatch *State of the World* report	
1985 Learning Alliance	1985 Toxic leak, Institute, West Virginia	
Rainforest Action Network	*Rainbow Warrior* blown up by French government, New Zealand	
	Antarctic ozone hole discovered	
	FDA approved bovine somatotropin	
	Dwellers in the Land: The Bioregional Vision (Sale)	
	1986 Nuclear plant explosion, Chernobyl, Ukraine	1986 Safe Drinking Water Act amended

		Superfund amended (Community Right-to-Know Act)
1987 Conservation International	1987 *Our Common Future* (Brundtland Report)	1987 Montreal CFC Protocol
	First debt-for-nature swap	Clean Water Act amended
	First U.S. Green meeting, Amherst	National Appliance Energy Conservation Act
		Marine Plastic Pollution Control Act
1988 Student Environmental Action Coalition	1988 Summer of the endangered earth	1988 Alternative Motor Fuels Act
	EDF-McDonald's agreement	
	1989 Earth as "Planet of the Year" (*Time*)	
	Exxon Valdez oil spill, Alaska (11 m. gallons)	
	Valdez Principles, Coalition for Environmentally Responsible Economics	
5. 1988–92	1990 Redwood Summer, California (Earth First!)	1990 Coastal Zone Management Act amended
	Dolphin-safe tuna agreement	Clean Air Act amended
	Earth Day, April 22	Northern Forest Appropriation Act
	"Big Green" proposition defeated, California	

TIMELINE

EVENTS

1991 Persian Gulf war pollution
disasters

Moralia (Mexico)
Declaration

Global Assembly of Women
and the Environment and
World Women's Congress,
Miami

1992 UNEP "Earth Summit"
conference, Rio de Janeiro,
Brazil

Anti-Columbus
Quincentenary
demonstrations

Record-high ozone-
destroying chemicals
detected, Northern
Hemisphere

World population at 5.5
billion

LEGISLATION

1991 Arctic National Wildlife
Refuge bill ended by threat
of filibuster

THE GREEN REVOLUTION

ORIGINS

IN the late summer of 1962, a marine biologist named Rachel Carson, already a best-selling author of books about ocean life, sounded this tocsin to the world:

> As man proceeds toward his announced goal of the conquest of nature, he has written a depressing record of destruction, directed not only against the earth he inhabits but against the life that shares it with him. The history of the recent centuries has its black passages—the slaughter of the buffalo on the western plains, the massacre of the shorebirds by the market gunners, the near-extinction of the egrets for their plumage. Now, to these and others like them, we are adding a new chapter and a new kind of havoc—the direct killing of birds, mammals, fishes, and indeed practically every form of wildlife by chemical insecticides indiscriminately sprayed on the land. . . . The question is whether any civilization can wage [such] relentless war on life without destroying itself, and without losing the right to be called civilized.

With those angry and uncompromising words, it can be said that the modern environmental movement began.

Silent Spring was an immediate, emphatic success, on the *New York Times* best-seller list for thirty-one weeks and with hardcover sales of more than half a million copies, rare for a serious nonfiction book. Its detailed and powerful condemnation of the American pesticide industry, particularly its startling case against DDT, touched a chord—a chord clearly waiting to be touched —made up in equal parts of a growing uneasiness with the world of postwar technology and a growing awareness of the nonmaterial amenities of life. Viciously attacked by the chemical industry, which mounted a $250,000 campaign to prove Carson a "hysterical fool," the book eventually won awards from the National Wildlife Federation and the Audubon Society, caused a wide popular outcry against the overuse of pesticides, and led directly to a Presidential Scientific Advisory Committee report in 1963 that corroborated Carson's work and reaffirmed her criticism of the pesticide interests; ultimately, it played a part in gaining support for the Pesticide Control Act of 1972 and the Toxic Substance Control Act of 1976.

But more than that: it galvanized a constituency no one had realized was there, energizing the somewhat sluggish traditional conservation groups as well as many who had never given a thought to the natural world before. Max Nicholson, head of the British Nature Conservancy and a figure of international renown, called it "probably the greatest and most effective single contribution" to "informing public opinion on the true nature and significance of ecology." American historian Stephen Fox put it even more succinctly: "The *Uncle Tom's Cabin* of modern environmentalism."

In 1960, Rachel Carson had discovered that a lump in her breast was malignant, and an operation failed to remove the cancerous growth entirely. Even as she was writing her book, her health grew steadily worse, despite therapy—"my body falters," she wrote a friend, "and I know there is little time left"—and in the spring of 1964, at the age of fifty-six, a victim of the very poisons she so skillfully analyzed, she died. "Man, alone of all forms of life," she had said in the chapter she had added on environmental carcinogens, "can create cancer-producing substances. . . . Human exposures to cancer-producing chemicals (including pesticides) are uncontrolled and they are multiple." And prophetically: "We tolerate cancer-causing agents in our environment at our peril."

In one sense, of course, it is misleading to date the American environmental movement from the summer of 1962, for an active concern for the natural world and alarm at its various perils actually go back through the twentieth century into the nineteenth (and even, with some few individuals like Thomas Jefferson, before). Artists and writers of the Romantic and Transcendental movements in the first half of the nineteenth century laid a firm foundation of appreciation for America's spectacular sights, a sensitivity built upon in the second half by a series of naturalists and activists as different as John Muir (a founder of the Sierra Club in 1892) and Gifford Pinchot (first head of the U.S. Forest Service in 1905). In the face of the impact on natural systems by increasing industrialism and commercialism, including such agencies of the federal government as the Army Corps of Engineers and the Forest Service itself, the twentieth century spawned a dozen environmental organizations (Audubon Society, 1905; Izaak Walton League, 1922; Wilderness Society, 1935; National Wildlife Federation, 1936) and several heroic environmental champions, Aldo Leopold, Joseph Wood Krutch, Rosalie Edge, and William O. Douglas prominent among them. And in the years after World War II, men like Howard Zahniser of the Wilderness Society and David Brower and Ansel Adams of the Sierra Club became notable public defenders of the wilderness, particularly in battles to prevent new dams in western rivers. Books like Fairfield Osborn's *Our Plundered Planet* and William Vogt's *Road to Survival* were best-sellers in the 1940s, and by 1960 membership in the major conservation organizations came to more than 300,000.

But, in another sense, it is fair to say that there was really no such thing as an environmental *movement*—concerted, populous, vocal, influential, active—before the publication of *Silent Spring*. As late as 1959, a speaker at the North American Wildlife Conference maintained that "the conservation conscience" would never be a conviction of the majority of the American people; three decades later polls showed that 80 percent of Americans supported the goals of environmentalism. A tendency had become a tide; old organizations had taken on new life and new ones sprang up on all sides; a new sensibility had in fact been born. Britain's Prince Philip, a longtime conservationist of the old school, called it, in some wonder, "the environmental revolution."

Such a revolution was not created by a single book, to be sure, however influential. The forces that lay behind the movement were multiple and complex. America was fifteen years into its postwar boom and just beginning to wonder how successful was "the affluent society" (as John Kenneth Galbraith named it in 1958) and at what cost it had been purchased. All around were the fruits of what has been called "the synthetic revolution"— plastics, fibers, chemicals, pesticides, detergents, nuclear power, and the like—and all around were the tracts and developments and suburbs and high-rises that fed on it, and yet it seemed not to have brought the comfort and harmony and tranquillity it had promised. Not only were vast (usually darker) segments of the population now seen to be somehow untouched by the benefits of the plastic highlife—John Kennedy's victory in 1960 was at least in part due to concern that "a quarter of the nation's children" were going to bed hungry—but those who had those benefits still felt a sense of crisis, as the social fabric frayed and modern diseases of "affluenza"—alcoholism, drugs, suicide, in-sanity, violence, alienation—increased. What's worse, all the material benefits apparently came at a price—urban crowding, suburban sprawl, pollution and smog, clear cuts and dammed rivers, cancer and nuclear fallout (a price increasingly displayed on television screens across the nation)—and came at a dizzying pace ("future shock," it was called) that seemed beyond the effective control either of the businesses benefiting from it or of the governments supposedly regulating it. In short, the discontents, disjunctions, and discords that were to become the troubles of the sixties were brewing.

At the same time, the postwar material boom had also produced a growing number of the college-trained, white-collar middle class that increasingly populated suburbs in search of trees and birdsong and stars at night. They wanted—and could afford—to concern themselves more with what the sociologists were calling "quality of life" (QOL) than with "standard of living" (SOL) in the traditional sense: the amenities beyond the necessities, including leisure time, outdoor recreation, healthy air and water, personal health and security, and hence a greater emphasis on the natural world, at least as represented by parks, preserves, wilderness areas, forest reserves, botanical gardens, and scenic highways.

Hence, when *Silent Spring* appeared, it found a ready audience, and that audience a cause.

That both audience and cause would continue to grow over the next thirty years probably could not have been predicted, given the way that so many social issues attract the attention of the capricious American public for a moment and then fade into obscurity. But environmentalism, as the new movement quickly became known, not only lasted beyond the particular furor over *Silent Spring* and the particular threat of DDT but continued to enlarge its scope, increase its concerns, expand its numbers and impact with every passing year—and with every new environmental disaster, of which there were many. By now, after thirty years, it has become an essential and indelible factor in everything from political campaigns to legislative agendas, building codes to product marketing, school curriculums to university research; in short, a permanent part of American life. Whatever shape environmentalism may take in the next decades, the next century, the movement of the last thirty years has altered American consciousness and American behavior, with consequences as profound as any movement since that against slavery in the nineteenth century. Rarely has a movement in so short a time gained such popular support, had such legislative and regulatory impact, produced so many active organizations, or become so embedded in a culture: a green revolution, indeed.

This book is the story of those three decades of the environmental movement in the United States, its causes and campaigns, it turns and twists, its heroes and enemies, its successes and failures. It is an era that can, roughly speaking, be divided into four coherent periods:

- The first, from 1962 to the first Earth Day in 1970, the time in which environmentalism was originally forced onto the nation's agenda, changing from the traditional conservationist concern for wilderness areas to a new focus on threats to human settlements.
- The second, coinciding with the decade of the 1970s until the Reagan presidency, during which Washington became the chief battleground and legal reformism the main effort, guided by a dark new perception of approaching doomsday and the human as an endangered species.
- The third, the years of the Reagan reaction, in which the movement divided into an increasingly professional mainstream and an increasingly radical grass roots, marked by a growing understanding of the earth itself as somehow endangered.
- The last, during Bush's administration, when environmental-

ism matured in both methods and impact and grew to new heights of membership and money, with both stubborn resistance and bitter frustration in the face of the awareness of stress and fragility and crisis in the biosphere of the only living planet in the known universe.

Three decades in the extraordinary experiment, unique in the modern world, of how humans might yet succeed in becoming guardians of that sacred vessel, Earth.

SIXTIES SEEDTIME

1962–70

WHEN, on April 22, 1970, a hundred thousand people thronged Fifth Avenue, closed to traffic because New York City officially recognized a celebration that called itself Earth Day, and ten thousand chanting, fist-waving demonstrators filled Union Square to cheer (and jeer) Gaylord Nelson, the United States senator who had thought the whole thing up, the strange marriage of 1960s rebellion and 1960s environmentalism was neatly symbolized. It was never an easy marriage, but, appropriately, both sides learned something and gained something, and when they made their affair public, as they did by the millions on Earth Day 1970, it was an impressive and percussive event.

Much of what is said about the sixties is misremembered nostalgia and most of the generalizations are bad history, but there was, from the North Carolina student sit-ins of 1960 to the sanctioned murders of four students at Kent State in 1970, a measurable and occasionally influential spirit of dissent and protest abroad in America. The environmental movement was by no means synonymous with that spirit—indeed, many of the long-established conservation organizations looked upon it with horror—but it was nonetheless affected by it, in some ways irrevocably so.

The protest generation of the sixties was a distinct minority, but it could resonate with such power through the society because it was challenging assumptions, expressing disillusions, and asking questions that in some way reflected the hidden doubts and fears of the great majority. Not all of them cared much about the issues of the environment until very late into the decade, but the sweeping nature of their dissidence, their willingness to oppose the consensus of the day, was very often infectious both for the mainstream environmental groups (at least among portions of their memberships, if not in their boardrooms) and for various ad hoc activist groups that sprang up. It would have been hard for any environmentalist in those heady days, especially one politicized, not to have been impressed by the boldness and directness of the challenge of the protest movement, its very immediate impact on the media (and thus often on society at large), and even its occasional successes.

This whole mode of political action, too, it must be remembered—sit-ins, demonstrations, protest marches—was new to this country, exemplifying a new style of citizen activism. Coupled with new rulings opening up public access to the courts and to governmental (and, increasingly, corporate) records, this led naturally to a myriad of new tools for environmental and public-interest groups, enabling them to engage an ever-larger grass-roots membership in an ever-wider assortment of tactics.

At the same time, the failures of midcentury society were increasingly underscored, and the consensus of the uncritical Eisenhower years fell apart. Whatever the nature of the failures—the foolhardy war in Vietnam, violence in the cities, assassinations and riots, a permanent "other America," devaluation and inflation—they showed that many of the hallowed systems of the land were in disarray. For many in the environmental movement, this meant an increased awareness of government as an environmental culprit—from malfeasance on the federal level to misfeasance on the state and nonfeasance on the local levels—and as a handmaiden of private business interests degrading the environment for private gain. Looked at closely, most of the agencies of the state—the Nuclear Regulatory and Federal Power commissions, the Army Corps of Engineers, the Agriculture and Interior departments— seemed driven more by economics than by ecology; "government," said academic researcher Lytton K. Caldwell, "emerged more often than not as a partner, promoter, or protector of activities that diminished the quality of the environment."

Then, too, the protest generation itself became the essential support base of environmentalism, joining the old-line conservationists first in trickles, then in waves. "We were just sitting here," one veteran of the National Wildlife Federation reported later, "and suddenly there they were, knocking at the doors." The "graduates" of the sixties supplied not only the shock troops for political demonstrations and anti-corporate actions but a good many of the publicists, lawyers, scientists, and other professionals in the civil rights, antiwar, feminist, and labor movements. This led to the growing influence of a younger set in most organizations, providing not only new ideas and styles but new tactics, ranging from innovative methods of raising money and lobbying Congress to cutting down billboards and other forms of what was called "direct action." And it led to the creation of any number of new environmental organizations, most of them local rather than national (Chesapeake Bay Foundation, 1966; Friends of the Everglades, 1969) but covering a wide range of concerns, including population growth (Population Crisis Committee, 1965), riverine protection (Freshwater Foundation, 1968), nuclear power (Union of Concerned Scientists, 1969), and ecological technology (New Alchemy Institute, 1968).

Thus energized and motivated, the environmental movement in the wake of *Silent Spring* took on a new life over the course of the decade. Its battle was fought on two main fronts, one traditional and the other new, in both cases with tactics both traditional and new. But now for the first time the impact was felt throughout the society, and now environmentalism moved from being the concern of the affluent and elderly of the boardroom on the one hand or of the backwoods hunters and fishers on the other, to being the stuff of everyday life—and politics—for millions.

The first front was the saving of the wilderness that had been the traditional impetus of the conservation movement since the turn of the century, and it was naturally the prime province of the traditional organizations, though aided now by the activist spirit in the land.

The fundamental attitude toward nature brought to these shores by the European colonists, and played out by every generation since, was that of utility and exploitation, but from early times, too, there had been a current of awe and appreciation, sometimes even of celebration. The history of the conservation movement until the mid-twentieth century was the conflict of this duality, as a small minority of celebrants sought to hold off the exploiters—John Muir called them "these temple destroyers, devotees of ravaging commercialism"—and the fact that some victories had been won in preserving wilderness was due to their efforts. But it had always been a negative kind of fight, impeding the extractors' course and staying the developers' hand, reacting to threats both corporate and governmental. The dream of taking the offensive, of forcing Congress to designate vast sections of unspoiled land as permanent wilderness territory—a dream going as far back as the 1920s, when Aldo Leopold persuaded the Forest Service to establish the first official wilderness area in Gila National Forest—lay unrealized for decades, fiercely resisted. Then came the sixties.

Howard Zahniser of the Wilderness Society had fashioned a wilderness bill for congressional consideration in the late 1950s, and hearings had in fact begun in 1957 and continued at a desultory pace in succeeding years, with powerful private interests and their legislators resisting every step of the way. But there was new energy in the conservationists' campaign in the new decade, and a new sense of how to go up against the recalcitrant government: traditional conservation groups like the Wilderness Society, the Sierra Club, the Audubon Society, and even the Izaak Walton League established a network to coordinate strategy and memberships, allies were found in such groups as the National Jaycees and the Federation of Women's Clubs, a stream of articles was prepared for both environmental and general magazines, and a drive was launched to generate vigorous grass-roots support. By 1962, Congress was getting more mail on the wilderness bill than on any other piece of legislation, and that was pressure that could not be ignored.

Eventually, as is sometimes the way of our national legislature, a wilderness bill came to a vote in the House in 1963 and the Senate in 1964, and it was passed overwhelmingly and signed into law on September 9, 1964. As is also the way, the act was substantially gutted: mining was permitted on wilderness lands well into the future, the President was given power to authorize dams and power plants in wilderness areas, only 9.1 million acres instead of the 60 million available in public lands (and 250 million in Bureau of Land Management lands) were designated, and a cumbersome system of bureaucratic and congressional control replaced a scheme of citizen-government cooperation. And yet it *was* a Wilderness Act, and it put the nation foursquarely in the business of forever preserving substantial areas "where the earth and its community of life are untrammeled by man, where man himself is a visitor who does not remain," in the words of its preamble; and in the twenty-five years after its passage, the National Wilderness Preservation System grew to more than 90 million acres comprising 474 units of public land. The Wilderness Society hailed it on its silver

anniversary as "one of the most important—even revolu-
tionary—conservation ideas in history."

Having flexed their muscles on the wilderness bill (and found
them stronger than ever before, if not as powerful as wanted),
the mainstream organizations again went into action when the
Bureau of Reclamation in 1966 announced plans to build two
dams to flood 150 miles of the Colorado River in the Grand
Canyon. The leader this time was the Sierra Club, drawing on
its heritage as the oldest conservation organization in the country, and, after a few somnolent gentlemen's-club decades, once
again regarded as an activist group with serious clout. And
although it was conceded that this was another of those defensive
and reactive battles of the kind that conservationists had strug-
gled through all along—building dams seemed to be a national
policy as powerful as the rivers themselves—this time the Sierra
Club brought into the fray a new man, a new strategy, and a
new constituency.

The man was David Brower, then fifty-four, and a thirty-year
veteran of Sierra Club struggles (and its executive director since
1952), who was enough in tune with the sixties to have put the
Sierra Club name forcefully in the consciousness of environmentalists and developers alike. He established a Sierra Club
publishing program in 1960 that brought handsome and dramatic photographs and descriptions of wilderness to millions of
chair-bound Americans, with total sales of over $10 million in
just that decade. And he personally traveled the country—and
the world—drumming up audiences at the drop of a cause
and delighting them with his sincerity and good humor. John
McPhee in a *New Yorker* profile called him the "archdruid" of
environmentalism and "unshakably the most powerful voice in
the conservation movement in this country."

Brower took on the Grand Canyon dam issue vigorously, and
with an array of tactics perfectly in tune with the times. First
he launched a series of full-page newspaper ads against the

projects ("This time it's the Grand Canyon they want to flood. *The Grand Canyon*"), complete with coupons and instructions on how to write the appropriate congressmen, a new device at the time; he then added pamphlets and bumper stickers, reprinted a stunning Sierra Club picture book, *Time and the River Flowing*, and ordered two full-color movies to be made; and he himself was indefatigable in making speeches and presenting testimony to the appropriate committees. It was the Sierra Club at the peak of its form, "the gangbusters of the conservation movement," as *The New York Times* labeled it in 1967.

And it worked. Mail flooded into congressional offices, thousands of phone calls and telegrams poured in, articles made the national press, and a network of grass-roots groups formed to keep the pressure on. (The government did nothing to help its case when it sicced the Internal Revenue Service on the club in June, denying its tax-exempt status on the grounds that it was engaged in "substantial" lobbying, which immediately aroused considerable public ire.) The Bureau of Reclamation offered a compromise of a single dam, a proposal roundly derided, and by early 1967 the Administration abandoned the projects entirely; Congress eventually agreed, formally dropping the scheme in July 1968. "Hell hath no fury," one congressman declared, "like a conservationist aroused."*

But what was aroused was something more than that, something that would play an essential part in the ongoing momentum of the environmental movement. For the first time, a large segment of the American public was heard to say that it preferred rivers to dams—dams, which had historically been favored by all administrations, by both parties in Congress, by

---

* Sierra Club membership shot from 39,000 in June 1966 to 67,000 in October 1968 (and 113,000 in 1970), making it the second-largest environmental organization (second to the giant National Wildlife Federation, 365,000 in 1968); but its loss of tax-exempt status meant the loss of its big donors and some severe financial problems.

developers of every stripe, and by generations who applauded turning darkness to dawn. Those dams were now seen by many as illegitimate concrete intrusions into wilderness areas that had their own integrity, their own beauty, and their own rights. The defeat of the Grand Canyon dams, together with the creation of the Wilderness Preservation System (and, later in 1968, a National Wild and Scenic Rivers System), marked a significant turning point in American attitudes toward the land: nature, or at least some remote parts of it, was not there simply for manipulation and exploitation but, other things being equal, should be preserved and protected and cherished.

The second principal front of the environmental movement in this decade, the one laid out squarely by Rachel Carson and made an important part of the American agenda for the first time, was generally given the broad heading "pollution," though in fact it was rather a dawning awareness of the dangers of a full array of human technologies to human health and safety.

The origins of this new sensibility probably lay in the fears aroused by the threat of nuclear fallout from atmospheric bomb testing and, to a lesser extent, the potential danger of radioactivity released from nuclear reactors. Some public anxiety was allayed by the cessation of aboveground tests as a result of the Nuclear Test Ban Treaty of 1963—regarded as the first international environmental agreement, even if the fallout issue was clearly secondary—but a nagging perception of the ancillary results of modern technology, the "side effects" as it were, persisted. And together with the disclosures of *Silent Spring*, showing the unanticipated evils of the "wonder chemicals" of the fifties, this created a nervous (if still confused) public that was highly sensitive to subsequent revelations of environmental dangers. Hence, when some eighty people died in New York City during an air "inversion" in the summer of 1966 . . . when the *Torrey Canyon* foundered and spilled 117,000 tons of crude oil into the English Channel in March 1967 . . . when an offshore oil rig near Santa Barbara, California, poured undetermined millions of gallons of oil along the California coastline in January and February 1969 . . . and when the Cuyahoga River near Cleveland burst into flames and the nearby Lake Erie was declared a "dying sinkhole" as a result of sewage and chemicals in the summer of 1969 . . . the public outcry was loud and widespread. Pollution, an apparently inevitable cost of business-as-usual, was no longer acceptable, and an increasing number of people for the first time announced themselves, with foot and pen and mouth, against the businesses producing it and the governments failing to protect against it. As the President's Council on Environmental Quality analyzed it some years later, referring to the Santa Barbara spill:

> No one was killed . . . no one suffered permanent health damage . . . no large numbers of people were threatened. . . . Yet the event dramatized what many people saw as thoughtless insensitivity and lack of concern on the part of government and business to an issue that had become deeply important to them. It brought home to a great many Americans a feeling that protection of their environment would not simply happen, but required their active support and involvement.

To some extent, the existing environmental groups could provide a channel for that involvement: the Audubon Society, for example, was an enthusiastic supporter of the anti-DDT campaign, the National Wildlife Federation in the mid-sixties began a legal challenge to polluters, and even the conservative, sportsman-minded Izaak Walton League was involved in various clean-water proposals. But none of the old-line organizations had the experience or expertise—or necessarily the desire—to act on the broad range of pollution issues, so it was left to a new set of smaller organizations to champion the citizens' concerns.

One such was the Committee for Nuclear Information, started at Washington University by Barry Commoner and some of his colleagues, which first took up the antinuclear cudgels and gradually moved to cover a broad range of pollution issues, mixing advocacy and science. Commoner was a biologist and a socialist, a far remove from the patrician mountaineers characteristic of the old mainline organizations, and his message was a mixture of his allegiances: first, the need to challenge corporations and the government on their versions of the facts; next, to propose more efficient and nonpolluting variations of their technologies; and finally, to try to take that technology out of the hands of capitalist corporations and into those of a democratic central government. It was a message he delivered skillfully, and increasingly often as the decade wore on, in books and articles and speeches and debates, and *Time* magazine soon recognized him in a cover story as "the Paul Revere of Ecology."*

Another group of new organizations was set up by an equally successful public champion, Ralph Nader, though his call was not for the abolition but for the reform of existing institutions. Starting first with the issue of auto safety in the best-selling *Unsafe at Any Speed* in 1966, Nader went on to respond to each new ripple in the activist wave of the sixties as if he was born for it: he established a Center for Study of Responsive Law in Washington, D.C., in 1968 to prepare lawsuits against a variety of corporations and their complicit regulators, then unleashed a series of task forces—Nader's Raiders—to study and recom-

---

* Commoner was only one of the many academics now drawn into environmental politics—prominent among them Anne and Paul Ehrlich of Stanford, Eugene Odum of Georgia, Raymond Dasmann of UC–Santa Cruz, LaMont Cole of Cornell, Kenneth Boulding of Michigan, and Garrett Hardin of UC–Santa Barbara—suggesting the measure of seriousness with which it was taken in the usually somnolent university world.

mend changes in auto safety, pesticides, nuclear energy, food and drugs, air and water pollution, with scathing exposés of the federal agencies supposedly in charge. In 1970 Nader started the Public Interest Research Group in Washington, devoted to action on those and similar causes, and spawned a series of similar organizations, largely with campus support, in almost half the states of the Union; the next year he set up an all-purpose lobbying and litigating center called Public Citizen, which by the end of the next decade had grown to some 50,000 members.

None of the new activist organizations was more ingenious than the Environmental Defense Fund, which was an important pioneer on the legal side of environmentalism, born as late as 1965 in the wake of a successful defense by the Sierra Club to block a power project on Storm King Mountain on the Hudson River. Established in New York City in 1967, EDF developed a staff of both ecologists and lawyers that very quickly developed the art of environmental litigation across a very broad front, from industrial pollution to nuclear-plant construction to lead poisoning, and proved that such tactics as lawsuits and injunctions (or the threat of them) were often more effective in a litigious society than letter-writing or lobbying. It also advanced the concept of environmental *rights*, a similarly attractive idea in a legalistic society, and EDF's Victor Yannacone argued that they applied not only to the relation of people to the natural world (the constitutional guarantee of "the right to a salubrious environment") but to nature itself, where, it was held, even trees had legal standing. On the model of the EDF, the Natural Resources Defense Council was set up with the help of the Ford Foundation in 1970 and the Sierra Club Legal Defense Fund was formed in San Francisco in 1971, and from then on, the "sue the bastards" technique was a fixed environmentalist strategy.

Shortly after EDF was founded, and also with Sierra Club

leadership, another group, Zero Population Growth, was formed in Washington. It was the outgrowth of the success of a book the club had published in 1968, Paul Ehrlich's *The Population Bomb*, which forced the issue of global overpopulation on the public consciousness in a fierce and apocalyptic way ("The battle to feed all humanity is over. At this late date nothing can prevent a substantial increase in the world death rate") and turned out to be the most popular environmental book ever published, with 3 million copies sold in the first decade. When in the resulting storm of controversy the book was attacked as being neo-Malthusian (which it was) and neo-Luddite (which it wasn't), Ehrlich's supporters at the Sierra Club, along with others in the environmental movement, decided to set up this lobby-cum-think tank to restate its message regularly and keep the issue alive.

The Sierra Club was also responsible for one other important new organization, though this time inadvertently. Brower's somewhat reckless handling of the club's finances, particularly in the wake of its loss of tax exemption and of its big-donor support, caused smoldering differences over his ideological emphases and confrontational style to flare into a palace revolt, and in the summer of 1969 the board denied him power over the club's purse. The beleaguered executive director took his case to the membership, but the annual board election returned a slate of anti-Browerites and Brower immediately resigned— with a certain bitterness on both sides, naturally enough, but with enough loyalty both ways so that Brower was named an honorary vice president of the club a few years later. Cut loose, Brower immediately launched a new organization he called Friends of the Earth, with a core staff in San Francisco and branches in both London and Paris, and a tax-deductible arm, the John Muir Institute for Environmental Studies, to avoid tax complications. FOE immediately established itself as the leading opponent of nuclear power plants and would go on to be the chief exponent of Amory Lovins and his "soft path" alternatives to national energy policies as well as a prominent voice on various pollution issues.

Energized by its era and creative in its response, the environmental movement by the end of the sixties represented a quite surprisingly popular cause, the full range of "pollution" issues now added to the traditional wilderness ones. Membership in the leading organizations, with the single exception of the somewhat staid Izaak Walton League, increased markedly, on a scale never before seen in the century:

Audubon Society	1962: 41,000	1970: 81,500
Izaak Walton League	1966: 52,600	1970: 53,600
National Wildlife Federation	1966: 271,900	1970: 540,000
Sierra Club	1959: 20,000	1970: 113,000
Wilderness Society	1964: 27,000	1970: 54,000

Mainstream media finally awoke to the issue and coverage gradually increased throughout the decade, reaching a crescendo in early 1970 with a spate of front-page articles and cover stories in *Time, Fortune, Newsweek, Life, Look, The New York Times*, and *The Washington Post*, typified by *Newsweek*'s headline, "The Ravaged Environment." "Ecology" became a word known to (if rarely understood by) the average citizen,* and arcane phrases like "environmental costs," "resource depletion," "atmospheric inversion," and "riverine eutrophication" were becoming common currency. "Ecology has become the Thing," economist Robert Heilbroner noted in *The New York Review of Books* in April 1970. "The ecology issue has assumed the dimensions of a vast popular fad," he said, though it was a good

---

* Despite headline writers, "ecology" is not the same thing as "environment": the environment is what is out there; ecology is how we study it, specifically the relation of species to each other and to their environment.

deal more than that: "The ecology issue is not only of primary and lasting importance, but . . . may indeed constitute the most dangerous and difficult challenge that humanity has ever faced." To have not only isolated that challenge but begun a response to it was an accomplishment that gave evidence that a social revolution in America had indeed begun.

The high point of that phenomenon was an event called, with typical sixties grandeur, Earth Day 1970. It had begun with a speech by Wisconsin senator Gaylord Nelson in Seattle in 1969 in which he proposed a kind of nationwide "teach-in" on college campuses, following the model of the antiwar teach-ins earlier in the decade. The response, Nelson reported, "was nothing short of remarkable," and so many inquiries poured into his office that he siphoned a $125,000 grant from some federal well and set up a staff of three Harvard graduate students in a Washington headquarters to handle the load. The Nixon White House was opposed, but enough political pressure built so that a number of federal agencies, particularly the Interior Department, were encouraged to help prepare materials for the event. Campus activists, then at a peak in opposition to an increasingly unpopular war, geared up enthusiastically to make the connection between the institutions supporting military intervention on the one hand and chemical pollution on the other.

No very precise calculation of the number of participants is possible, but April 22 saw probably the largest of all the demonstrations of the 1960s. According to the organizers, 1,500 colleges and 10,000 schools took part, many campuses had street demonstrations and parades, and large rallies were held in New York, Washington, and San Francisco. *Time* estimated that some 20 million people were part of what Nelson called "truly an astonishing grass-roots explosion": "The people cared and Earth Day became the first opportunity they ever had to join in a nationwide demonstration and send a big message to the politicians—a message to tell them to wake up and *do* something."

There were some who denounced Earth Day as a Communist plot—the Daughters of the American Revolution, for example, declared it was "subversive"—but it is a measure of sixties radicalism that a greater number protested it as an Establishment plot, an attempt to get the attention of youths and students away from the war and civil rights and back to something that could be discussed in "reasonable" circles. Certainly the fact that many mainstream politicians gave their support, including a number of Nixon cabinet officers, suggested that environmentalism was more acceptable than Black Power or antiwar opposition, and the fact that certain conservative foundations like Resources for the Future and the Conservation Foundation provided finances suggested that its message was not thought to be all that subversive. *Ramparts*, a short-lived left-wing California magazine, devoted a special issue to Earth Day finances, arguing that its intention was not a true preservation of the earth but merely "a more efficient rape of resources."

But whether or not it could muster mainstream support, Earth Day 1970 was in fact a surprising demonstration of the depth of feeling about environmentalism at that time. *The New Republic* probably put that understanding best: Earth Day was "not just a channel for frustrated antiwar energies as we had thought"—before the affair, it had dismissed environmentalists as "the biggest assortment of ill-matched allies since the Crusades"—but rather an event that "signaled an awakening to the dangers in a dictatorship of technology," by people in all walks of life. Or, as *Audubon* put it with some pride: "Now, suddenly, everybody is a conservationist."

As Gaylord Nelson realized, a demonstration of that magnitude had to have a political effect in Washington. The response of Congress had been slow and piecemeal throughout the sixties, and extremely gingerly where large economic interests were concerned, and so the few measures it had taken—a Clean Water Act in 1960, a Clean Air Act in 1963, and a Solid Waste Act in 1965—were quite cautious and limited. Now, however, a citizens' trumpet had been sounded and the politicians had to show that they heard it. The Clean Air Act was amended to give it a few more teeth, a Water Quality Control Act was passed to improve drinking water, a Resource Recovery Act regulated some toxic waste disposal, and a National Industrial Pollution Control Council was established. In all this, of course, the clear purpose was to respond to the public without jeopardizing the status quo so revered by federal legislators—without, for example, forcing automobile manufacturers to adopt stringent emission-control standards, or banning heavy metals from sewage-treatment streams, or restraining the output of untested chemicals—but it was unquestionably a beginning, a camel's nose under the tent.

The act that was to be the most influential of all—that, to the surprise even of its supporters, was to be the substructure for all the environmental reform legislation in the following decade—was the National Environmental Policy Act of 1969. (Signed into law on January 1, 1970, by a reluctant Richard Nixon, it actually predated Earth Day but gained its support from the same forces.) This sweeping law created the enormous Environmental Protection Agency—very soon to become the largest regulatory body, in both people and budget, ever established in the government—which consolidated responsibility for enforcing and shaping the myriad of federal laws and regulations passed over the years, and a Presidential Council on Environmental Quality, which was assigned to oversee the nation's environmental health and prepare an annual report. The act also gave birth to a new idea that eventually created nothing less than a new industry: the environmental impact statement. Originally designed to force all government agencies to forecast the consequences of any future project on policy "significantly affecting the quality of the human environment" —later broadened to include local governments and most non-governmental projects as well—it thus assured that a parade of scientists, academics, lawyers, engineers, and bureaucrats would find employment for years on end. (No fewer than 12,000 federal EISs were prepared in the next ten years.) The federal government, as ill-prepared and unwilling as it might have been, was at last irrevocably in the environmental management business.

In its first eight years the environmental movement compiled a record of accomplishment that no one could have calculated when it began, but even more important, it achieved a broadened scope that no one could have imagined. The concern was no longer just the impact of human society on the wilderness and its species; it now included the impact of human society on humanity as well. And though some important victories would be won on the former front in the decade to come, it was primarily the latter that was the preoccupation of movement and public alike.

What was at work here was not simply a shift in concerns or a reordering of priorities but rather, in some deep way few at the time could articulate, a profound change in philosophical groundings. The earlier understanding had been more or less *biocentric*, holding that the human species was only one among many and that all creatures had legitimate (and some would say equal) rights that should be respected: wilderness areas should be largely inviolate, forests left alone (or cut selectively, at most), rivers allowed to run free, animals left wild and hunted with care. Thus, the protection of wild lands, though at times it would require appeals to backpacking and other human use, had essentially a biocentric motive: as David Brower put it, "I believe in wilderness for itself alone, I believe in the rights of creatures other than man." The newer understanding (though indeed its roots go deep into Western culture) was for the most part *anthropocentric*, on the one hand regarding forests as a source of timber, rivers as a source of electric power, animals as a source of food and power, and on the other believing in the sanctity and health of the human habitat above all others. Thus, the issue of pollution became primary, concerned as it was with the protection of the human species and all it had to touch and ingest: as *Audubon* put it, "There is a new battle to be waged—to keep man's technology in check while promoting the welfare of man himself."

These essentially contrary points of view would become more antithetical, and heated, as time went on, but for the moment at any rate they served more to empower and enlarge the movement than to divide and limit it. As it went into the next decade, it had every reason to feel confident, what with its newfound successes and its newfound popularity, and above all its newfound sense that the environmental game could be played best of all in Washington, where a vigilant national government could be made to see to the wide and prudent management of national resources as well as the fair and healthful regulation of dangerous industries. It didn't quite work out that way, of course, but at least the challenge was in the air and the game afoot.

(3)

DOOMSDAY DECADE

1970–80

IN the icy middle of January 1972, the British magazine *The Ecologist* published a special issue called *A Blueprint for Survival*, at once a scathing attack on industrial society and its environmental perils and a detailed plan for its short-term amelioration and long-term replacement. Based on a wealth of new scientific and economic data then beginning to be assembled in the West, it was the outcome of a conference put together by the magazine's editor, Edward Goldsmith, and modeled on a similar gathering sponsored by the Massachusetts Institute of Technology two years before.

But at heart it was not a technical document at all. It was a straightforward prediction of doomsday—"if current trends are allowed to persist, the breakdown of society and the irreversible disruption of the life-support systems on this planet, possibly by the end of the century, certainly within the lifetimes of our children, are inevitable"—and an impassioned plea for the developed nations, with Britain in the lead, to establish an ecological society in place of the failing industrial one. It was signed by a distinguished set of prominent scholars, among them Frank Fraser Darling, Julian Huxley, Peter Medawar, E. J. Mishan, and C. H. Waddington, and as a paperback became an instant best-seller in Britain.

A sampler: "The principal defect of the industrial way of life with its ethos of expansion is that it is not sustainable. Its termination within the lifetime of someone born today is inevitable. . . . Radical change is both necessary and inevitable because the present increases in human numbers and *per capita* consumption, by disrupting ecosystems and depleting resources, are undermining the very foundations of survival. . . . It should go without saying that the world cannot accommodate this continued increase in ecological demand. *Indefinite* growth of whatever type cannot be sustained by *finite* resources. This is the nub of the environmental predicament."

The doomsday chord was played often in the decade of the seventies as the world awoke to its environmental crisis, and if it was sometimes sounded louder than any facts seemed to warrant, it was nonetheless a vivid recognition of the new condition of the world. Among the books whose notes contributed—and their titles tell the story—were Samuel Mines's *The Last Days of Mankind*, John Loraine's *The Death of Tomorrow*, Ron Linton's *Terracide*, John Maddox's *The Doomsday Syndrome*, L. S. Stavrianos's *The Coming Dark Age*, Richard Falk's *This Endangered Planet*, Anne and Paul Ehrlich's *The End of Affluence*, Donella Meadows's *The Limits to Growth*, and Gordon Rattray Taylor's *The Doomsday Book*, and that's just a sampling. Most, like *Blueprint*, were done by professionals, often scientists, and most were filled with studies and statistics as well as anecdotal and sometimes computer-modeled evidence. Which did not mean that they were necessarily reliable—indeed, not a few of the direst predictions turned out to be exaggerated—but they did have a convincing and sobering impact on press and public alike.

And the doomsday chord indeed had plenty of real-life accompaniments just then to give it convincing substance. Among them: the OPEC oil embargo of 1973 and the resulting gasoline shortage and doubling of prices, when the idea of finite resources hit home for the first time; the publication in 1974 of studies showing the danger of chlorofluorocarbons (CFCs) to the ozone layer, and the subsequent 1976 National Academy of Science report condemning aerosol sprays; the mysterious death in 1974 of Karen Silkwood, a whistle-blower exposing dangers at the Kerr-McGee plutonium plant in Oklahoma; the North Sea oil blowout in 1977 in which 8.2 million gallons of crude gushed into the water and along shorelines; the accident at the Three Mile Island nuclear power station in March 1978, the first acknowledged release of radioactivity by a domestic power plant; the disclosure in 1978 that the soil on which the homes and school along Love Canal in Buffalo, New York, were built was contaminated with eighty-two different chemicals, some known to be toxic, leading to a $27 million evacuation and buyout; the grounding of the *Amoco Cadiz* tanker in 1978 off the coast of Bretagne, France, spilling some 70 million gallons, the worst oil disaster to that date; and the report by the EPA in 1979 that there were between 32,000 and 50,000 major hazardous waste sites in the United States, at least 2,000 of which were regarded as posing significant risks.

To this was added a growing awareness of a host of other environmental problems which, though not necessarily bursting into the news with catastrophe or crisis, were studied and reported on as never before: groundwater contamination, loss of topsoil, deforestation, filling of wetlands, acid rain, global warming, nuclear waste disposal, species extinction, polluted fisheries, toxic smog, elimination of landfills, overhunting of whales, ozone depletion, and on and on, seemingly without letup. Only the very blind and the willfully ignorant could fail to realize at last the seriousness of human risk to human and other planetary life.

The kind of movement that evolved in these years and the kind of public support that sustained it were both largely without precedent, at least in this century. For one thing, the interest groups that sprang up around environmentalism were in the largest sense *disinterested*, serving a general citizens' purpose for the most part, not tied to traditional parties or power groups and dependent on true, broad public support for the largest amount of their incomes and political clout. For another, the sheer number of such groups seeking social change in the 1970s, small and large, national and neighborhood, was well beyond anything seen before, growing from perhaps several hundred as the decade began to an estimated 3,000 at its end, ranging from a dozen prime movers with national memberships to a myriad of kitchen-table and church-basement operations coast to coast. And then, too, the range of interests and causes (not to mention styles and strategies) was absolutely protean, diverse, and varied as no other movement to date, as is suggested by these few examples:

- Environmental Action, which grew out of the successful Earth Day demonstrations in 1970, became one of the most aggressive of the Washington-based groups—"tough, able, and persistent" is how one writer described their lobbyists—and highly visible especially on matters of toxic waste, nuclear power, and energy policy.
- Greenpeace, begun in 1971 largely to protest nuclear testing, first broadened into a widely popular campaign to "Save the Whales" and then sea lions, dolphins, and other oceanic species, and by the end of the decade added a full range of issues, including tropical forests and toxic-waste traffic.
- Environmentalists for Full Employment, started by Richard Grossman and a few others in 1975, was an effort to confront the claim corporations were increasingly making that restrictions on pollution and waste would cost jobs—a bugaboo that carried considerable weight with some segments of the work force—by showing that environmental cleanup and restoration

would improve workers' health on and off the job and could actually increase employment to boot.

- Worldwatch Institute, primarily a research and publication center, was begun by Lester Brown in 1975 to gather information worldwide on global threats, which has been published in more than a hundred pamphlets and in an annual *State of the World* report widely respected in the media.
- Not to mention the Center for Marine Conservation (1972), Citizens for a Better Environment (1970), the Cousteau Society (1973), Earthwatch (1971), Food First (1975), the Hunger Project (1977), the Institute for Local Self-Reliance (1974), the Jane Goodall Wildlife Institute (1971), Negative Population Growth (1972), Sea Shepherd (1977), and the Trust for Public Land (1972), counting just national organizations.

As important as the multiplication of new organizations was the increased strength and visibility of the old ones, which continued to grow both in membership and in muscle. The top five established organizations showed a combined increase in membership from 841,000 in 1970 to 1,485,000 in 1980, a growth of more than 76 percent, but perhaps more revealing is that it was the three most active and visible groups and the ones that broadened from the traditional wilderness constituency to the new pollution-minded audience that gained the most, while the old-style groups went nowhere:

			%
Sierra Club	1970: 113,000	1980: 165,000	+46
National Wildlife Federation	1970: 540,000	1980: 818,000	+51
Audubon Society	1970: 120,000	1980: 400,000	+330
Wilderness Society	1970: 54,000	1980: 50,000	−7
Izaak Walton League	1970: 53,000	1980: 52,000	−2

Among the other established organizations—Defenders of Wildlife (1947), National Parks and Conservation Association (1919), and the Nature Conservancy (1951)—membership gains were also reported, though the staffs of these organizations tended not to favor much the sixties-influenced activism of their confreres.

By the middle of the decade a combination of old and new organizations, based in Washington or with Washington offices, had come together into a visible power center in the capital, not so much a formal coalition as a loose alliance, kindred spirits who could be counted on to join forces for a number of the important legislative battles of the day and whose chief officers would meet periodically. At the core was a circle that soon became called the "Group of Ten" (or even the "Big Ten"), regarded as the most important "players" on the capital scene: Audubon, Defenders of Wildlife, Environmental Defense Fund, Environmental Policy Institute, Izaak Walton, National Wildlife, Natural Resources Defense Council, National Parks Association, Sierra Club, and Wilderness Society. (And, in the immediate periphery, Environmental Action, Friends of the Earth, League of Conservation Voters, Nature Conservancy, and World Wildlife Fund.)

The game for these players and their colleagues, however, was a traditional Washington game, ranging from effective constituency pressure, vote producing, and committee testimony on the one hand to wining-dining, old-boy-network favors, and back-room trading on the other. To play it meant their taking on at least some of the coloration of the other high-powered players in town. Lobbying and public pressure organizations—Audubon, National Wildlife, Sierra Club among them—found it important to recruit scientists with backgrounds in formal ecology and economists with some experience in social cost accounting, plus experienced Capitol Hill hands for lobbying, professional fund-raisers for foundation support, and mail-order specialists for expanding memberships. Litigation-centered organizations, too—EDF, NRDC, Sierra Club, LDF in particular—found it necessary to recruit top-flight legal talent as they found themselves involved with congressional law-making and tangling with federal bureaucrats hiding behind regulatory tape. And both kinds of groups turned increasingly to academics and MBAs and technospecialists of all kinds able to compete in a high-powered and high-stakes world with the staffs of Senate committees and well-heeled corporate legal offices and industry association lobbyists and high-ranking Administration appointees.

And thus, and quite quickly too, the major environmental organizations began to turn into more professional and down-to-business operations than at any time in the past, predictably with mixed consequences. Though they certainly played an ever-larger part in what became the "industry" of environmental legislation and regulation, and with practical measures on Capitol Hill and in federal courts and through EPA directives to show for it, it was true that some of the earlier passions and commitments began to fade: in the words of historian Samuel Hays, "The Washington-based environmental movement of the 1970s emphasized practical gains rather than affirmation of ideologies." And it was also true that organizations of this kind did little to address the problem of an increasing distance from two constituencies historically part of social-change coalitions: organized labor (in such extractive sectors as logging and mining and such regulation-threatened industries as automobiles and chemicals) and urban blacks (in communities where environmentalism was seen as a luxury of those who could afford to go backpacking in the wilderness or move up to more efficient cars).

What Congress unleashed by creating the Environmental Protection Agency in 1970 was more than just an agency that grew faster and farther than any brought into being before, though that growth was impressive: starting with a staff of 6,000 and a budget of $455 million in 1971, the EPA grew in just a decade to nearly 13,000 people with a budget of $5.6 billion and the power to distribute another $1.6 billion in toxic-waste cleanup funds. (This made it one of the largest federal agencies; only the Veterans Administration and the Office of Personnel Management were bigger.) It also was, as is the way with modern megagovernments, the favored dumping ground for almost all the awkward, persistent, nagging problems of pollution and human health that kept ending up on legislators' desks. Between 1970 and 1980, Congress passed no fewer than eighteen far-ranging and complex environmental acts, no small feat for a branch of government noted for its snail-paced responses to even the most urgent matters.

The breadth of that legislation was quite impressive: after the five major acts of 1970, a Water Pollution Control Act and the Pesticide Control Act (finally banning the domestic use of DDT, ten years after Carson's book, though permitting manufacture for export) in 1972; a Safe Drinking Water Act in 1974; a Toxic Substances Control Act (theoretically to oversee the introduction of new chemicals) and a Resource Conservation and Recovery Act (to control hazardous wastes) in 1976; amendments to both the Clean Air and Clean Water acts, providing additional teeth and broader coverage, and amendments to the Ocean Dumping Act to bar sewage sludge in 1977; an Energy Policy and Conservation Act, belatedly responding to the 1973 energy crisis, in 1978; and in 1980, in the last months of the Carter presidency, a spate of laws including regulation of low-level radioactive waste, research into alternative motor fuels, and especially creation of a Superfund (Comprehensive Environmental Response Act) to identify the most dangerous toxic sites across the nation, and to force and finance their speedy cleanup.

Nor was that all. Though pollution issues were by now clearly most urgent, Congress also responded to a range of more general wilderness problems. The most important single achievement was the passage in 1973 of the Endangered Species Act, which gave the Fish and Wildlife Service power to determine which animal and plant species were threatened with extinction (and how severely so) and to take certain modest measures for their protection and restoration; the rationale was that such species have "aesthetic, ecological, educational, historical, recreational, and scientific value to the Nation and its people," about as anthropocentric a justification as a wilderness act could have. Other notable responses included several to protect animal life (Marine Mammal Protection Act, 1972; Fishery Conservation Act, 1976; and Fish and Wildlife Conservation Act, 1980) and others to protect habitat (Coastal Zone Management Act, 1972; Federal Land Policy and Management and National Forest Management Acts, 1976; and Alaska National Interest Lands Act, 1980). And the wilderness lobbies showed their power once again—and, as always, the power of outraged local citizens behind them—in forcing Congress in 1975 to abandon two projects—the Hells Canyon dam on the Snake River in the West and the Tocks Island dam on the Delaware in the East—that would have intruded hydroelectric projects into free-flowing rivers.

There were significant deficiencies in most of this legislation, to be sure. Congress's response was inevitably patchwork reformist at best, and time after time it forced environmental groups, no matter what their politics, to accept that stance: we will tinker, it said, we may even regulate, but we will not fundamentally alter the economic or political systems that produce the perils we are asked to confront.

Assumptions about the necessity, the inviolability, of economic growth, for example, underlay virtually every congressional decision and were not about to be flouted no matter what the environmental contingency. Establishing clean air standards was one thing, and Congress was eventually able to figure out how to do it, but actually taking the serious steps necessary to achieve them—forcing Detroit to reduce auto emissions, say, or shutting down coal-burning utility plants—proved impossible to do. Adding new acres to the national forests system was possible from time to time—though only after considerable pressure and persuasion—but not without provisions that the land was to be open to corporate timber cutting (including clear-cutting) and oil and gas drilling, with only 16 percent of the area designated as wilderness and that small amount precarious.

Moreover, Congress was always readier to regulate than eliminate: no matter what the risk, it preferred to have some agency like the EPA formulate and monitor "safe" levels and "acceptable" amounts rather than banning or discontinuing some product or practice understood as dangerous. As a legislative tactic it was often an effective way to win over needed support, particularly from industry-minded lawmakers, and to convince constituents that action was being taken, but as an administrative tool it was usually ineffective and sometimes counterproductive. The congressional ban on DDT, for example, though a decade tardy, was largely successful in preventing domestic manufacture and use because it mandated simple and straightforward prohibition, easily enforceable and generally acceptable. Congressional action on dioxin, though, was quite useless because it provided only for corporate good behavior and bureaucratic oversight, and that was only one of many thousands of chemicals that the EPA has been asked to "regulate," limiting source production or setting acceptable risk limits or mandating control devices, almost always with little perceivable effect.

Finally, the only perspective Congress could ever muster was a narrow anthropocentric one. Only the Endangered Species Act had any pretense of having been passed on behalf of other than two-footed creatures (and even then benefits to "the Nation and its people" came first), and all of the wilderness preservation legislation, even when it chose to deter commercial development, presumed that such areas were for human use and pleasure, even for eventual human medical scavenging. Whatever the rhetoric that might be attached to the various acts, the guiding principle, as set forth in a U.S. Forest Service proviso in 1978, was that "wilderness is not set aside for the sake of its flora and fauna, but for people."

Nevertheless, all that granted, the congressional achievement should not be gainsaid. The Congresses of the seventies irrevocably made the federal government an essential participant in the multifold processes of control, evaluation, prohibition, and regulation of the full range of industrial and commercial activities, from extraction to disposal, in the United States. Pollution became the people's business and the wild lands and the species within them continued to be part of the people's protected heritage. Under pressure from a growing environmental movement, Congress evolved something often called a conservation consensus—more appropriately seen as an environmental entente—that coalesced frequently and successfully enough to take the important first steps to solidify environmentalism on the public policy level, with ripple effects throughout subsidiary governments and the society as a whole.

The United States, though well among the leaders in responding to the perceived dangers of the doomsday scenarios, was by no means alone. By 1971 at least eleven other nations had environmental regulatory agencies in place, and by the end of the decade there would be nearly a hundred more; at the same time more than 400 independent private global organizations, both scientific and political, had formed in response.

The first significant international recognition of the environmental crisis was actually a private one, the formation of a worldwide coalition of scientists, politicians, technocrats, and businessmen following a meeting in Rome in 1968 under the guidance of Aurelio Peccei, an Italian entrepreneur. By 1970 this Club of Rome had seventy-five prominent members in twenty-five countries, mostly in the industrial world, and was prepared to launch the project that ultimately emerged in March 1972 as *The Limits to Growth*, a complex computer-modeled analysis of global economic and environmental trends. It concluded, to considerable fanfare in the world press, that the cause of environmental degradation was exponential growth of the global industrial machine and that severe restrictions on population, resource extraction, and agricultural expansion were necessary to avoid a catastrophe by the year 2000. Attacked at once by both economists professionally committed to growth-at-all-costs optimism and social scientists professionally critical of what was called "computer fetishism," the *Limits to Growth* argument nonetheless forcefully made what was essentially the "enlightened capitalist" case—scale back now so as to keep industrialism itself going for the future—and it enjoyed a worldwide impact, with eventually 4 million copies sold in thirty languages.

It was just about this time that the official international response emerged, and in a form quite similar to that of the Club of Rome. The first global meeting was the Biosphere Conference in Paris in 1968, sponsored by the United Nations Educational, Scientific, and Cultural Organization, and though it went little beyond a recognition that environmental threats were "producing concern," it did succeed in prompting the United Nations to put together a full-scale effort in Stockholm in the summer of 1972. That UN Conference on the Human Environment proved to be a serious and remarkably cohesive event that announced now for the first time that ecological problems were to be taken seriously by the nations of the world—or, as Barbara Ward, the British economist, put it to the opening session of the conference, it was "one of those

turning points in man's affairs when the human race begins to
see itself and its concerns from a new angle of vision." In
practical terms, it passed, without dissent from any of the 113
national delegations assembled, a broad declaration proclaiming
"the urgent desire of the peoples of the whole world" to assure
"the protection and improvement of the human environment"
—again it is the *human* condition that mattered most—and it
derived a list of twenty-six general environmental and economic
principles and an "Action Plan" intended to carry them out,
principally through a monitoring body called Earthwatch. It
also endorsed a continuing UN environmental body, to be called
the UN Environment Programme, which eventually established
its headquarters in Nairobi, Kenya, with a staff of nearly a
thousand. And it inspired the creation of several thousand "non-
governmental organizations," national and international bodies
of various sizes and seriousness, devoted to environmental
matters in the loosest sense, and fostered an atmosphere in
which, over the next decade, a number of multinational and
binational agreements could be forged for environmental remediation and protection.*

The Stockholm conference was a notable achievement, considering how new and complex the issue was, but its limitations
were marked, and they indicate the profound difficulties that
would continue to confront the governments of the world,
collectively and individually, in the coming decades. For one
thing, the interests of the less-industrialized nations—the so-
called Third World—were not by any means identical to those
of the industrial world, and the fundamental issue of how the
poorer nations were to rush pell-mell into industrialization
without experiencing all the difficulties of the richer nations was
simply pushed aside; indeed, the conference ended up declaring

* In the forty years since 1930, forty-eight international conventions or treaties
on the environment had been signed; in the years from 1971 to 1980, forty-
seven were agreed to.

that nations have "the sovereign right to exploit their own resources" and that environmental policies should "not adversely affect" development in the Third World. Similarly, the conference was unwilling to take up the broader issue posed by industrial societies anywhere or to consider whether industrial systems by their very nature might be incompatible with stable and renewable ecosystems; instead, the entire thrust of the meeting was in terms of "stewardship" of the earth, managing it successfully so as to ensure future economic growth both North and South with minimal environmental costs. Lastly, with its single-minded focus on the "human environment" and earth entities as "resources," it gave no attention to wilderness (that indeed was dismissed as a peculiarly American concern) or species diversity, endangered habitats, ecological restoration, or human overpopulation.*

Stockholm's essential message—that human misuse of the biosphere had reached crisis proportions and all nations must respond—was repeated, with various additions and glosses, at a series of successive world gatherings, in 1974, 1976, 1977, and 1981, all under the auspices of UNEP. Unfortunately, the mechanism for getting nations to respond—UNEP itself—was an example of international bureaucratization at its worst, painful evidence that neither individual states nor collective bodies were prepared to alter seriously the long-established systems that led to the acknowledged crisis. UNEP was set up, not as a specialized agency like UNESCO or the Food and Agriculture Organization with its own programs and powers, but as a coordinating body that had to work through other

---

* It is not conspiratorial to note that Maurice Strong, the UN official who was secretary-general of the conference and later UNEP's executive director, was a Canadian multimillionaire businessman with a background in resource extraction, or that the intellectual groundwork for the conference and its commissioned report (*Only One Earth*, by Barbara Ward and René Dubos) was laid by an International Institute for Environmental Affairs financed and led by Robert O. Anderson, chairman of the Atlantic Richfield petroleum company.

agencies, most of which resented and resisted its interference. It was given a paltry budget—$20 to $30 million a year for its first decade—and told to coordinate all matters on global ecosystems without any powers of coercion or enforcement. And it was headquartered in Nairobi, far from decison centers either corporate or governmental, where the influence of Third World nations played a disproportionate role in decision making, or the lack of it. Still committed to traditional forms of industrial development—typified by huge dam projects complete with hydroelectric plants and ecosystem destruction—and suspicious that environmental restrictions were meant to hold back legitimate processes of modernization, these governments (or at least their political elites) resisted any significant UNEP influence in their affairs, and discouraged its actions elsewhere.

Little wonder, then, that for its first decade UNEP made only the most sporadic progress in its Earthwatch network and regional-seas protection programs, could muster support only for such tepid treaties as the ones designed to protect Antarctic seals and to prohibit trade in endangered species, and in such responsibilities as slowing desertification and establishing disaster-warning systems it was a sorry failure. Indeed, ten years after Stockholm, UNEP's own full-scale reassessment claimed only "a mixed record of achievement" and found many areas where, in bureaucratese, "progess has been very slow."

One last international response during these years can be seen as related to the inadequacy of either private or governmental responses: the growth of the Green Party movement. The roots of this movement can be traced to a Values Party started in New Zealand in 1972 and an Ecology Party in Britain in 1973, both of which were inspired to make environmental issues the cornerstone of a new kind of electoral strategy, and Green organizing on this model took place in at least a dozen other countries during the decade. The core of the Green idea was that public policies affecting the environment in the broadest

sense—from resource extraction to population dispersion, nuclear armaments to foreign aid—were too important to be left
to conventional politicians mired in pre-environmentalist mind-
sets. Success with such a sweeping platform was not readily
forthcoming, but Green politics became increasingly attractive
to at least an activist minority—especially in Europe, where legal
gains such as had been won in the United States were scarce—
and Swiss Greens won seats in parliament in 1979, Green parties
contested elections in France, Britain, Luxembourg, and West
Germany that same year, and in 1980 West German activists
formed Die Grünen, destined to become the most important
Green manifestation of the 1980s.

There was no Green movement yet in the United States by the
end of the seventies, but environmentalism, working on other
fronts, had made itself an accepted fact of public life. A poll
taken in 1980 by Resources for the Future found that 7 percent
of the American people described themselves as being "environmentally active"—that would translate to some 15 million
individuals—and another 55 percent said they were sympathetic
with the aims of the movement; another poll, by *The New York
Times*, determined that as many as 45 percent felt that "protecting
the environment is so important that requirements and standards cannot be too high and continuing environmental im-
provements must be made regardless of the cost." Such an
expression of support for a movement that was not yet twenty
years old was quite remarkable in the annals of American
politics.

There were deficiencies in this support, to be sure. Generally
the strength came from the more urbanized and often more
liberal sections of the country—chiefly the conurbations known
as Bos-Wash, Chi-Cin, Pitt-Phil, Sea-Port, and San-San, the last
in California—and petered out in the Heartland and Plains,
dwindling even more in such areas dependent on petrochemicals

and extraction as the Rockies and the Gulf. Generally the shock troops as well as the directors and the staffs were younger, richer, and better educated than the American average, leading to criticism that it was an "elitist" movement concerned only with the luxuries of a good life for those who already had a good living. And generally both the organizations and their supporters were predominantly white, with little impact in black or brown populations where ecological matters at this point seemed trivial compared with economic ones and the connections between the two rarely made. (Black sentiment of the time was fairly summed up by Whitney Young of the Urban League: "The war on pollution is one that should be waged after the war on poverty is won.")

Nonetheless, it was nothing short of remarkable that, in an era of increasingly large and remote governments normally unresponsive to popular concerns, the environmental movement had been able to mobilize a large segment of the public and have its voice quite frequently heard in legislative and regulatory chambers. Remarkable, too, that it had been able to translate this into a series of apparently indelible laws and permanent institutions with wide social and economic implications and impacts. And most remarkable that it had achieved this on behalf of a set of aesthetic, hygienic, and ecological values that had hitherto had insignificant weight in American life, and in the face of powerful corporate, bureaucratic, and special-interest resistance.

A Blueprint for Survival, in its typical doomsday fashion, had argued that "if man wishes to survive, to ensure the proper functioning of the self-regulating mechanisms of the ecosphere must be his most basic endeavour." Whatever the shortcomings of the American movement, it could be said at least to have raised this ultimate issue and started to force its society to face up to its implications.

4

THE REAGAN REACTION

1 9 8 0 – 8 8

I N the summer of 1980 a document entitled *The Global 2000 Report to the President*, written by the White House's Council on Environmental Quality and a task force from the State Department, was placed on the Oval Office desk of President Jimmy Carter. It was not a comforting report. Not only was the peril facing the world as a result of overpopulation, environmental abuse, and resource depletion real and threatening, it said, but there was "the potential for global problems of alarming proportions by the year 2000." Worse still, "given the urgency, scope, and complexity of the challenges before us, the efforts now underway around the world fall far short of what is needed."

It is of some note that eighteen years after *Silent Spring*, ten years after Earth Day and the creation of the EPA, an official document could be quite so gloomy: not only had the problems of the biosphere not been eased after attention had been drawn to them but they seem to have gotten worse, and perhaps were even beyond comprehensive solution. For therein lay the awful persistent paradox at the heart of environmentalism, one that came to be understood with increasing force as the eighties evolved: the fight to save the earth, however popular and successful by all conventional norms, was failing to save it in any real sense, seemed in fact to be uncovering only new threats and crises.

Yes, certainly, there were solid achievements. Laws had been passed, court cases won, agencies created. Lands and trees and waters had been saved from developers, some befouled rivers were clean again, some threatened species were protected, some widespread toxics were banned. Limits on worldwide whaling catches were established, states were beginning to enact bottle-return and other recycling laws, nuclear power plants were placed under EPA regulation, the "fast breeder" reactor was killed. Millions of people were alert to the perils and willing to see some actions taken, even if—and this was confirmed in polls—the economy suffered and taxes were raised.

And yet *The Global 2000 Report* was uncompromising: "The earth's carrying capacity—the ability of biological systems to provide resources for human needs—is eroding," caused by "a progressive degradation and impoverishment of the earth's natural resource base. . . . If present trends continue, the world in 2000 will be more crowded, more polluted, less stable ecologically, and more vulnerable to disruption than the world we live in now. Serious stresses involving population, resources, and the environment are clearly visible ahead," among them a 50 percent increase in world population, a rise in world poverty, a decrease in vital supplies and healthy ecosystems, and a growing vulnerability to both natural and human catastrophes.

Six months after the delivery of *Global 2000*, on January 21, 1981, Ronald Reagan, a film actor and former governor of California, was inaugurated as President. The report was relegated to the nearest archive, neither its warnings nor its recommendations given the slightest heed, and the country that was by itself already the greatest single threat to global health embarked on a decade of unchecked speculative economic growth, fueled by an unprecedented increase in the national debt from $4 trillion to $11 trillion—those are figures with twelve zeros—that within two terms made the United States the largest debtor nation in the world, and in history. Just as the magnitude of the environmental peril was beginning to be understood and the need for serious remediation appreciated, the forces of the American Establishment chose to deny the evidence, ignore the warnings, and coalesce behind a champion of business-as-usual—or bigger-than-usual.

One might say it was like an ostrich burying its head in the sand at the sight of danger. Except that ostriches do not in fact behave that way—they put their beaks through sand in search of food, but run, and quickly, when threatened—only people do.

The Reagan Reaction that began the 1980s was a specific backlash against the environmental innovations of the 1970s and the resultant large regulatory system which, despite its inadequacies and inefficiencies, impinged significantly on corporate America and threatened to become only more burdensome in time. Right-wing businessmen like Richard Mellon Scaife and Joseph Coors, and conservative treasuries like the Mobil and Olin foundations, poured money into ad campaigns, lawsuits, elections, and books and articles protesting "Big Government" and "strangulation by regulation," blaming environmentalists for all the nation's ills from the energy crisis to the sexual revolution. Several well-heeled think tanks were established under this "New Right" aegis, the most prominent being the Heritage Foundation in Washington, which was soon using a $10 million-a-year budget to lay out a sweeping backlash agenda for the pro-business, pro-development administration it expected to bring to the capital in the 1980 election.

Ronald Reagan was the perfect advocate for exactly this conservative power bloc, and he rode into Washington, thanks to its immense power and his own considerable charm, with a mandate to "get government off our backs" and "set business free again." It was a task of such apparent simplicity that he and his advisers set about it with great enthusiasm, and his first years in office were mostly a fulfillment of the Heritage plan to dismantle government regulations and squelch environmental and other public-interest influence in Washington. What was not effected by judicious cabinet appointments was attacked with budget maneuvering or legal resistance or congressional arm twisting, and when all else failed, with a simple refusal to act and a quiet policy of nonenforcement.

The Office of Management and Budget, created by Richard Nixon to draw fiscal power to the White House, was used adroitly by Reagan's men to reorder government finances away from regulatory functions, especially environmental. The Council on Environmental Quality was deprived of half its budget and most of its staff, reduced to a nullity in a clear case of shooting the messenger with the bad news. Agencies like OSHA and EPA were eviscerated by budget cuts—EPA lost 29 percent of its budget and a quarter of its staff in the first two Reagan years—and innovative programs in areas like solar energy and alternative fuels were scrapped altogether. When such restraints were not enough, the White House let it be known that environmental restrictions for industries such as mining, timber, oil, and automobiles would be operationally ignored or unenforced; when industry forces objected to EPA directives, their trade associations were invited to participate in drawing up new regulations on terms they felt were satisfactory.

On top of that, the Reagan people quite cynically used their power of appointment to put individuals in charge of departments and agencies who would downplay existing regulations and resist new ones. Those put in charge of such key federal agencies as the Forest Service, the Bureau of Land Management, the National Parks Service, and OSHA were avowed enemies of government rules and restrictions. Robert Harris, for example, made the head of the new surface-mining agency Congress had created in 1977, had been the chief opponent of and instigator of a lawsuit against that very agency. Anne Gorsuch,

Reagan's choice to head the EPA, was a Colorado lawyer whose regular clients included many extractive and agricultural interests that had openly opposed federal regulation; of the fifteen subordinates she named to EPA positions, eleven had been connected with industries that the EPA regulated.

The most outrageous appointment was that of James Watt as Secretary of the Interior. Watt was a Colorado right-winger who had headed a Mountain State Legal Foundation set up in the late 1970s as a means by which large corporate interests in the West could fight back against environmentalism, funded by such millionaires as Joseph Coors; his general feelings were expressed in a 1983 speech in which he compared environmentalists to both Nazis and Bolsheviks. In charge of all the nation's public lands, he sought to increase commercial use of them by timber, mining, livestock, oil, tourist, and other industries; typical was his expansion of a program of "deficit sales" of timber lands to West Coast lumber companies, allowing them access to millions of acres below market value at taxpayer expense.

It was only a small act but it was telling: in 1982, the United Nations adopted a World Charter for Nature, a document proclaiming that "nature shall be respected and its essential processes shall not be impaired" and that "man's needs can be met only by ensuring the proper functioning of natural systems" and suchlike. Every single nation in the world body endorsed the charter but one. Under orders from Ronald Reagan, the United States voted in opposition.

For its first two years the Reagan Reaction was nearly unstoppable. Environmental legislation ground to a halt—the Nuclear Waste Policy Act of 1982 was the only significant new law between 1981 and 1986, and it was studded with pro-industry provisions—and environmental lobbies were frustrated time and again on Capitol Hill. A White House Task Force on Regulatory Relief, under Vice President George Bush, spared countless corporate polluters from government action and forced the EPA to delay or scrap regulations on workplace chemicals, hazardous wastes, and automobile emissions, among others. Offshore drilling increased, timber cutting expanded, federal lands were sold to private interests, wilderness areas were opened to oil and gas leasing, the endangered species list lay moribund, new chemicals came on the market without testing, and evidence of each new ecological crisis was greeted with "further research is necessary."

But it could not last. For all its undoubted power, the backlash simply could not halt the tide of history. The environmental crisis, underscored by repeated tragedies—the evacuation of dioxin-contaminated Times Beach, Missouri, in 1983, the poisoning of several hundred thousands (and some 3,000 deaths) at the Union Carbide pesticide plant in Bhopal, India, in 1984, the discovery of the ozone hole over Antarctica in 1985, and the disastrous explosion of the nuclear plant at Chernobyl, in the Ukraine, in 1986—was too real, too threatening, to be denied. And the environmental movement, still growing with new organizations and new public support in response to that crisis, was too entrenched, too necessary, to be thwarted. Indeed, if any further evidence was needed to demonstrate the importance of environmentalism, a 1985 Harris poll found that 80 percent of the public—about four times the number that had voted for Ronald Reagan—supported current environmental regulations and standards.

And the irony is that the environmental majors—the Group of Ten and their colleagues—actually ended up gaining from the Reagan assault.

Anne Gorsuch got herself mired in an EPA scandal over her failure to enforce Superfund cleanups and refusing to testify to Congress about it, and was forced to resign in March 1983; one of her deputies, Rita Lavelle, was jailed for six months in a related perjury. James Watt proved to be so abrasive and insensitive in both word and deed, and so openly scornful of the public he was supposed to serve, that he too became a political liability and resigned in the fall of 1983. Both furors resulted in new support for the environmental majors—"We're sorry to see him go," one lobbyist said of Watt, "he was the best organizer we ever had"—and some estimates put their total membership at more than 5 million by mid-decade. The Sierra Club showed the most spectacular gains, partly as a result of an aggressive membership drive, going from its 1980 level of 165,000 to more than 350,000 by 1985, but many of the others also grew by mid-decade: the Audubon Society to 450,000, the National Wildlife Federation to 825,000, the Wilderness Society to 100,000, and Friends of the Earth to 25,000.

The Reagan challenge also inspired the movement to develop new tactics and methods. Electoral work was emphasized more than ever before, and in the 1984 elections environmental organizations played a part in at least a third of the congressional races and numerous state contests. The League of Conservation Voters (originally formed by David Brower out of Friends of the Earth in 1970 but now a separate D.C. organization) and Environmental Action both came up with lists of the "Dirty Dozen," legislators with anti-environmental records selected to be opposed for reelection. The Sierra Club and FOE among others established political action committees outside their lobbying work to solicit funds for direct electoral work, and both the Sierra Club and Audubon set up computerized indexes of their membership lists by congressional district to better muster local support for key congressional elections.

In spite of the signal lack of legislative success in the first Reagan term, Washington became increasingly the focus of attention for environmental players in the eighties; the number of registered environmental lobbyists in the capital rose from 2 in 1969 to 88 in 1985. Lobbying and research staffs—the latter more and more important as laws became more complex—were increased at most offices: the Sierra Club had 17 paid staff by 1986, Audubon over 200, and the most powerful of them all, the National Wildlife Federation, expanded to more than 500 in the capital alone. Feeling the lure, Friends of the Earth began shifting its emphasis from the San Francisco staff to its Washington office and in 1985 in an acrimonious split finally moved its headquarters to the capital, prompting founder David Brower to resign from the FOE board and complain that the organization was becoming "just another lobbying group." What was called "Potomac Fever" seemed to characterize much of the movement, and having success among the Washington power brokers— along with policies that would assure credibility, cooperation, and contacts on the Hill—became a top priority for most of the large organizations.

At the same time, the majors went through a not so subtle and perhaps inevitable process of institutionalization, settling into large buildings, large staffs, and large budgets. Most marked was the change in leadership in the middle of the decade: Audubon, Sierra Club, EDF, Greenpeace, Wilderness Society, and Defenders of Wildlife all installed new administrators, all of them managerial types such as lawyers, corporate executives, or bureaucrats, and all at six-figure salaries—in each case brought in from the outside and with a new emphasis on management, personnel skills, and budget balancing. "This is the big time," one Audubon staffer said, "and we've simply decided to become professional."

As a result of this maturing, the environmental establishment began to take on the coloration of its surroundings. "Both sides are wearing suits and lugging laptops now," one Washington veteran reported in early 1985, "and they both talk the same 'cost benefit' and 'social risk' eco-jargon, too." A *National Journal* ranking of the most effective lobbying organizations in Washington, those with the most clout with decision makers both legislative and administrative, listed Sierra Club, Wilderness

Society, and NWF among the top. (A survey of Washington policymakers at about the same time by *Environmental Forum* magazine ranked the most effective environmental groups, in order, as Natural Resources Defense Council, National Wildlife Federation, Conservation Foundation—affiliated with the World Wildlife Fund in 1985—EDF, Audubon, League of Women Voters, and Sierra Club.) As a symbol of the growing interchangeability, former government bureaucrats began to show up on the boards and staffs of the environmental organizations: Cecil Andrus, Carter's Secretary of the Interior, served as an Audubon director and Wilderness Society adviser in the early 1980s, and Terry Sopher, head of Interior's wilderness program, went to work as the Wilderness Society's chief lobbyist at the Bureau of Land Management.

The Washingtonization of the environmental majors was a mixed blessing. On the one hand, it did enhance the movement's influence with important legislators, particularly Democrats on a few key committees, and with some important administrators, particularly in the Interior Department and the EPA, where William Ruckelshaus replaced Anne Gorsuch. In Reagan's second term the clout paid off with amendments to stiffen the Safe Drinking Water Act (1986) and Clean Water Act (1987)— Gaylord Nelson, father of Earth Day and then chairman of the Wilderness Society, noted that the 1986 act allotted $18 billion for waste-water treatment, as against $250 million in the budget twenty years earlier—and especially with a series of amendments to the Superfund Act in 1986, one of which, the Community Right-to-Know Act, established once and for all that the citizenry could force both private industry and public agencies to disclose what kinds of pollutants they might be producing in any phase of their operations. In smaller ways, too—forcing changes in pork-barrel legislation, for example, cutting appropriations for road building in national forests, prodding for additions to the Endangered Species List, pressuring the Defense Department on nuclear-waste cleanups—the D.C. groups had a growing influence on government policy. "The environmental movement used to be about stopping things," said a senior attorney for the NRDC in 1986. "Increasingly, it's about doing things."

On the other hand, a kind of "inside the Beltway" camaraderie, the symbiosis of lobbyist and lawmaker, outside expert and inside expert, subtly changed the role of the environmentalists more and more from adversaries to allies. With both sides staffed by professionals, both seeing the need for compromise in the sloppy process that is modern American government, it was only natural that they should work for accommodation rather than confrontation—and if the accommodation meant, as it often did, taking crumbs today in hopes of bread tomorrow, well, that's the way the game is played in the big time, and the crumbs are, after all, big-time crumbs. "Sure," said one of the top men in the FOE office around this time, "it's often a matter of one step forward and two steps back. But if we weren't here, it'd be zero steps forward and six steps back, and sometimes you just have to fight anyway, even if it's a losing battle."

But there were consequences, and they did not go unnoticed in the ranks. When the majors put out a joint publication in late 1985 called *An Environmental Agenda for the Future*, the first common document to spell out national policies, they were immediately attacked for having produced a "top-down" book that ignored the advice of regional and local leaders in the field and for having become lost in a Washington tunnel cut off from "more far-seeing, more imaginative" ideas, as the *High County News* in Colorado put it, guilty of a "reflexive adherence to legislative solutions." And when even bills touted as major pieces of legislation failed to get out of committee or got seriously compromised on the floor, members and staffers alike grew bitter and restive. "There's too much movement now away from the ideals and too much emphasis on bottom lines," David Brower complained. "The MBAs are taking over from the people who have the dreams. Do MBAs dream?"

One example, perhaps not entirely typical, is revealing of the Washington tango. In 1981, responding to growing public pressure, mostly from the grass-roots organizations that had grown up around nuclear-power sites, Democratic congressman Morris K. Udall of Arizona, the doyen of environmental politicians in the House, fashioned a bill to deal with high-level radioactive waste, most of it spent reactor fuel. It created a means for finding and testing dangerous-waste areas and a provision for a permanent national waste repository by 1998, but it also gave the nuclear industry some significant protection from either citizen or state grievance procedures and some relief from hard-won EPA oversight, and it permitted a system of surface storage to reprocess spent fuel, benefits much sought by the pro-nuclear lobbies. Many veteran anti-nuclear activists were appalled by the bill—about two dozen local groups, including the Clamshell Alliance in New England and the Abalone Alliance in San Francisco, united to oppose it—but it had the significant support of the Sierra Club, Environmental Action, the Environmental Policy Center, and FOE, all of whom regarded Udall as one of their best friends in Congress and a person not to cross when he had a pet bill in the hopper. With that endorsement Udall could tell his colleagues his bill had the support of the serious environmental community (the Sierra Club eventually withdrew its support, but quietly and without effect) and the fact that there was no important opposition from the Group of Ten meant that the bill could easily pass the House and go on to become law.

As the environmental majors expanded their Washington operations, a process that would continue for the next decade, two other important dimensions were added to (or became more noteworthy in) the movement: grass-roots activism and radical environmentalism.

Some part of environmentalism had always been primarily local, simply because many of the problems—nuclear plants, waste dumps, factory emissions—were local. But with the eighties, and the growing feeling that official Washington was unresponsive and environmental Washington preoccupied, grass-roots organizations proliferated; Peter Borrelli, editor of the NRDC's *Amicus* magazine, estimated that some 25 million people were involved one way or another at the local level by 1987–88. With the passion of people whose lives were intimately affected and an energy fired by what came to be called the NIMBY (Not in My Backyard) syndrome, these groups made themselves heard by both state and city agencies and local corporations, often with telling effect. "Today the action is bottom-up," Borrelli noted, "since it is at the local level that laws and programs set in place over the last two decades are implemented"—or, just as often, not. It was just such action that led to the passage of Proposition 65 in California in 1985, an anti-toxic initiative against state agricultural and chemical industries, the first successful environmental initiative since 1972.

The grass-roots response was often much tougher and less compromising than those of national organizations, both because the local activists did not have large disparate constituencies to worry about and because they had, literally, to live with the decisions made. "If someone's worried about the health of their children," as one activist put it, "they won't be convinced by appeals to 'political pragmatism.' " Or as Barry Commoner saw it:

The older national environmental organizations in their Washington offices have taken the soft political road of negotiation, compromising with the corporations on the amount of pollution that is acceptable. The people living in the polluted communities have taken the hard political road of confrontation, demanding not that the dumping of hazardous waste be slowed down but that it be stopped.

Grass-roots organizations also had a broader reach and, in usually undeveloped ways, a somewhat deeper perception than the nationals tended to have. Minority groups of all kinds and many blue-collar neighborhoods were drawn to environmental activism out of some local need—particularly because they were often targets of undesirable and dangerous projects that affluent communities resisted—whereas the majors were made up largely of white and more affluent staffs and constituencies. Women, too, were disproportionately represented in both membership and leadership of local groups, often housewives with little previous activism but a number who were veterans of various protests of the sixties. And because such people were in the trenches, as it were, they tended to have much less reverence either for the assurances of officialdom or for the pronouncements of experts, all of which they treated with a healthy distrust, and they were much less inclined to believe in the inevitable worth of economic growth or the unquestioned right of corporations to make decisions affecting local social and environmental affairs.

The most impressive evidence of grass-roots power came with the hottest issue of the decade, toxic waste. Largely at the instigation of Lois Gibbs, a housewife whose effective leadership of the residents of the Love Canal neighborhood brought her national attention, a Citizens' Clearinghouse for Hazardous Wastes was formed in 1981 to coordinate and assist the work of local groups. By the fall of 1986 it had a network of 1,300 groups, two-thirds of them begun after 1984, when news of the evacuation of dioxin-infested Times Beach and the explosion at the chemical plant in Bhopal was prominent in the media; by the end of the decade it reported working with no fewer than 7,000. Organized around such issues as groundwater contamination from landfills, dumping of industrial chemicals and heavy metals, and new incinerators for municipal garbage, such groups energized many people who had been politically inactive and exerted their power with letter-writing campaigns, town meetings, door-to-door canvassing, and even demonstrations and civil disobedience. With encouragement, information, and advice from the Clearinghouse—for example, on up-to-the-minute alternatives for sewage treatment that could win over reluctant town boards—many of the locals were able to gain substantial concessions or outright victories, usually to the surprise of their high-powered antagonists. One indication of the alliance's effect is that since the Love Canal crisis in 1978, no new hazardous-waste dumps have been established in America. "Not because they're illegal," Lois Gibbs is careful to point out, "but because people have lobbied at the grass roots."

To their credit, the national organizations did not take long to respond to evidence of activism at the grass-roots level. Many of those with local branches—Sierra Club and Audubon especially—decided to intensify local organizing (though keeping effective decision making at the national level); Sierra even began a National Toxics Campaign to coordinate local actions primarily against military waste sites, spun off as a separate organization in 1984 with headquarters in Boston and a dozen offices around the country. And several national organizations made it a point to provide legal and technical help for local groups in need, as the NRDC did with more than three dozen suits in mid-decade against local polluters, including one on behalf of the grass-roots Chesapeake Bay Foundation against Bethlehem Steel which ended in the largest-ever settlement in a pollution case.

The second new dimension of the eighties, radical environmentalism, was similarly decentralized and often emerged in similar reaction to the nationals, but it was usually inspired by people with considerable political experience, much of it tinged by the insights of the sixties and often informed by years of work inside the mainstream movement. Their causes and their tactics, not to mention their styles and rhetoric, grew directly out of opposition to what they saw as the reformism and the "cooptation" of the mainstream at a time when the perils seemed to be multiplying and the national leadership unresponsive. Among the charges that they leveled was that the old organizations were too legalistic ("You should never support a piece of legislation," said Dave Foreman, radicalized after a decade of suit-and-tie lobbying in Washington, "you should always be asking for more"); too professional ("You've got a new group of bureaucratic professionals," asserted Lorna Salzman, a onetime FOE activist in New York, "who are not in it for a cause but because it's a 'public interest' highfalutin *job*"); and too limited ("The reform environmentalists have no program and no vision," argued George Sessions, a professor of philosophy at Sierra College, "they're about on the level of the penal establishment").

The emergence of this new breed and their criticism of the majors were serious enough to prompt Michael McClosky, director of the Sierra Club from 1969 to 1985 and its subsequent chairman, to send a confidential memo to his board of directors in January 1986 warning of the "new, more militant" environmentalists. "They are people who do not hesitate to criticize the main players such as the Sierra Club," he wrote, but their target is larger, to change "the relationship of individuals to society and the ways in which society works." The question they pose to the movement is "whether it is wise to work within the context of the basic social, political, and economic institutions to achieve stepwise progress, or whether prime energies must be directed at changing those institutions." And he added: "They're just utopian. We may be 'reformist' and all, but we know how to work within the context of the institutions of the society—and they're just blowing smoke."

Not quite smoke. The new radicals could sometimes be more vociferous than they were coherent, sometimes let frustrations lead them into actions insufficiently planned, sometimes were trapped into taking positions in public without having done enough homework—in short, showed the failings of any large group of disparate people acting in the public arena against the status quo. But in the decade of the eighties they made their mark.

Despite differences, sometimes substantial, what generally united the radical environmentalists was an underlying criticism of the dominant anthropocentric Western view of the world and a feeling that the transition to an ecological or biocentric view had to be made with all possible speed, with active and dramatic prodding if necessary. Such a sensibility was deeply ecological, in that it understood the true interdependence of species and their habitats (and the necessarily limited role of the human among them), and deeply radical too, in that it demanded a profound change in the values and beliefs of industrial society from the bottom up. Altogether, in the words of philosopher George Sessions, "it shows us that the basic assumptions upon which the modern urban-industrial edifice of Western culture rests are erroneous and highly dangerous. An ecologically harmonious paradigm shift is going to require a *total* reorientation of the thrust of Western culture."

Among the expressions of this new radicalism, four overlapping tendencies stand out.

Bioregionalism, the idea that the earth is to be understood as a series of life territories defined by topography and biota rather than by humans and their legislatures, was the first to take root in America. It imagined human societies organized on the lines of empowered bioregions, expressing such values as conservation and stability rather than exploitation and progress, cooperation and diversity rather than competition and uniformity, and decentralism and division rather than centralization and monoculture; as one early formulation put it, "the bioregional movement seeks to re-create a widely shared sense of regional identity founded upon a renewed critical awareness of and respect for the integrity of our natural ecological communities."

The movement itself began in California in the late 1970s and by the mid-1980s it encompassed some sixty local organizations: some were explicit bioregional councils, as in the Ozarks, the Kansas prairie, the Hudson Valley, and the Northwest; some, such as the National Water Center in Arkansas and Friends of the Trees in Washington State, had the specialized interests their names implied; some, including those in Appalachia, the Columbia River valley, the San Francisco Bay area, and Cape Cod, published regular magazines on bioregional themes. The first of a series of biannual continental congresses, designed to set policies on environmental issues and establish movement-wide links, was held in the Ozarks in 1984, since then followed by meetings in Michigan, British Columbia, Maine, and Texas.

Deep ecology, originally formulated by Norwegian philosopher Arne Naess in the seventies, was brought to the United States primarily by George Sessions and sociologist Bill Devall, who co-authored its first popular account in 1984. Standing in contrast to what Naess termed the "shallow environmentalism" of most of the movement, deep ecology stressed such points as: ecological equality, the right of every species to existence and survival and with equal "intrinsic value" regardless of its importance for humans; the diversity and abundance of all life forms, which should not be reduced by humans except "to satisfy vital needs"; the sharp reduction of human population so that other species may not only survive but have sufficient habitat to thrive; the preservation of the wilderness as a pristine habitat valuable in its own right; and the self-realization of humans through lower levels of consumption and resource use. Complicated as they were, such ideas quickly gained a following in the United States—and elsewhere in the world, including Canada, Australia, and Northern Europe—and proved especially influential among both radical activists and academic philosophers, no mean feat.

Deep ecology was subject to a savage attack by veteran writer Murray Bookchin and some of his followers, beginning with a biting diatribe at the first U.S. Green meeting in Amherst in 1987. Bookchin, a longtime anarchist in the old-left tradition and creator of what he called *social* ecology, accused deep ecologists of ignoring social systems and their injustices, and failing to see the "social roots" of the environmental crisis; scorning biocentrism as just another "centricity," he argued that it was degrading to think of humans as "a mere species" and "Malthusian" and "misanthropic" to advocate population reduction. The controversy, at different levels of heat and light, went on for several years and in many periodicals, but in the end seemed to come down to whether one saw the ecological crisis as the result of faulty psychosocial attitudes of industrial society (materialism, humanism, technophilia) failing to achieve a harmonious, spiritual relationship with nature, as the deep ecologists would argue, or whether one saw it as a result of faulty socioeconomic arrangements of capitalist society (class rule, hierarchy, bureaucracy) committed to an accumulative and competitive market relationship to nature, as the social ecologists would say. Either way, a crisis.

Ecofeminism, a synergistic blend of sixties-style feminism with eighties-style ecology, placed its emphasis on the connections between the domination and exploitation of women and the domination and exploitation of nature, both seen as products of a male-dominated society. Inspired in part by two books, Susan Griffin's *Woman and Nature* in 1978 and Carolyn Merchant's *The Death of Nature* in 1980, ecofeminism sought to go beyond the limits of earlier feminist ideologies, particularly by raising issues that set women in a context wider than just the economic. "Why is it that women and nature are associated, and vilified in our culture?" asked one early proponent, Ynestra King. "Does the liberation of one depend on the liberation of the other?" It also sought to go beyond what were seen as the limitations of other radicalisms by raising questions about "androcentrism," the male-focused perspective, as the real heart of the eco-crisis and about patriarchy as the central instrument in understanding the Western domination of nature. Like deep ecology, ecofeminism had a considerable following on the campuses, in women's studies and philosophy departments particularly, and inspired a veritable torrent of books and articles in this decade; several ecofeminist conferences were held in these years as well, the largest and most comprehensive at UCLA in the spring of 1987.

The Gaia hypothesis, formulated by British scientist James Lovelock in a small book in 1979, suggested that because the earth was apparently so regulated as to maintain its temperature, its atmosphere, and its hydrosphere with extraordinary precision for millions of years, it could in fact be thought of as a living organism. Immediately popular among many nonscientists as a useful metaphor for thinking about a biocentric earth, the Gaia idea spawned a number of similar analyses (as well as conferences, T-shirts, study groups, and an oceangoing Viking ship), all supporting positions congenial to the radical perspective. Interestingly, the hypothesis was seen to embody perceptions not very different from those of various early tribal peoples, including the American Indians, whose record as model ecologists was being brought to light at about this time; it was characteristic of most Indian mythologies to think of the earth as a single living being and to derive ways of behavior and thought that would ensure its careful, productive existence.

These expressions of radical environmentalism naturally gave rise to a great many organizations in these years, several of which had national importance. Among them:

• Earth First!, the more or less organized expression of the activist side of the new radicalism, was started by Dave

Foreman and a handful of other disillusioned operatives from mainstream environmentalism around a campfire in 1980. Designedly formless, without national staff, bylaws, formal incorporation, or even membership, it was simply dedicated to the principle that "in *any* decision consideration for the health of the earth must come first" and that in carrying this out, it should make "no compromise in defense of Mother Earth." Inspired in part by novelist Edward Abbey's 1975 *The Monkey Wrench Gang*, Earth First!ers stood foursquare in defense of wilderness and its biodiversity and made militance a cardinal part of their tactics, soon including guerrilla theater, media stunts, civil disobedience, and, unofficially, "ecotage" (also called "monkey wrenching"): sabotaging bulldozers and road-building equipment on public lands, pulling up survey stakes, cutting down billboards, destroying traps, and, famously, "spiking" trees at random to prevent their being cut and milled.* No sure way exists of checking such a figure, but an EF! spokesperson has said that the cost to the nation of such ecotage was $20–$25 million a year.

With such forthright militance, EF! succeeded in attracting a considerable following and by the end of the decade had grown to more than seventy-five chapters in twenty-four states (mostly in the Southwest and on the West Coast) and Mexico and Canada. But it paid a penalty for its success: as Foreman put it, "from one side there are concerted efforts to moderate us, mellow us out, and sanitize our vices; from another side have come efforts to make us radical in a traditional leftist sense; and there are ongoing efforts by the powers that be to

* The only known injury from tree spiking was to a millworker at a Louisiana Pacific mill in Cloverdale, California, in 1987, when a band saw struck an embedded spike, which the company blamed on EF! and the media broadcast with great intensity. EF! was never charged or even investigated, however, no evidence ever connected it to the spiking, and the tree was not in an old-growth area activists had been defending nor was it even standing when it was spiked, not a monkey-wrench tactic.

wipe us out entirely." Such pressures—including FBI infiltration and a trumped-up federal suit against Foreman and others in July 1989 and a car-bombing of two California activists in May 1990—eventually led to Foreman's dropping out and the group's splintering into several rival groups in the early 1990s.

- U.S. Greens, originally formed as the Committees of Correspondence at a Minneapolis meeting in 1984, took as their model the German Greens, who had had twenty-seven members elected to the Bundestag the year before and seemed to be showing the way for other Green parties in Europe and beyond. The American version expanded its philosophical "pillars" from the four of the German Greens (ecological wisdom, social responsibility, grass-roots democracy, and nonviolence) to include six others (community economics, decentralization, post-patriarchal values, diversity, global responsibility, sustainability) in a kind of New Left catchall designed to appeal to all constituencies. Fearing a "top-down" organizational shape, the original founders constructed an ungainly system of local chapters and regional representatives, plus a "clearinghouse" in Kansas City, and this was the none-too-effective vehicle for the Green movement in the United States for the next seven years.

Eventually the organization grew to more than 200 locals in all fifty states, changed its name to Green Committees of Correspondence (later just U.S. Greens), and embarked on the creation of a platform and governing body that would give it national shape, confirmed by a national convention in 1991. Wracked by factionalism, however, and by sharp disputes over tactics, particularly the issue of party politics and contesting elections, the Greens found more success at local levels, especially on the West Coast (it gained ballot status in California in 1992), than as a national presence.

- Sea Shepherd Conservation Society was started by Paul Watson after he was kicked out of Greenpeace for being too militant; it became the method by which he lived out a vision he had had during a Sioux sweat-lodge rite that he was destined to save the mammals of the ocean, especially whales. With a "navy" consisting of a single ship, he and his crew had dedicated themselves to being the police of the seas, eventually incapacitating at least seven vessels illegally hunting whales, confronting ships illegally fishing with gill nets that trap marine mammals and birds, and taking direct action, not excluding ecotage, to prevent seal hunts in Canada, dolphin slaughter in Japanese waters, and whaling in the North Atlantic. The organization, which has some 15,000 support members, has adopted a slogan of "We don't talk about problems, we act," and it has lived up to it.

• Planet Drum, the first formal expression of the bioregional movement, was established in San Francisco in the 1970s to be an active center for publications, speakers, performances, and workshops on this new philosophy. Through its biannual *Raise the Stakes* newspaper and a series of "bundles" from different bioregions that it distributes to its membership, it has become the effective networking core of the movement; its "Green City Program," designed to show how bioregional ideas can be effected in urban areas, has been used as a model for ecological platforms in San Francisco and other West Coast cities.

Though the range of radical environmental groups has been extremely wide, and not always cohesive, there is no doubt that it expressed a real current of the environmental movement, one that has had an impact primarily in local areas and on specific issues but has also prodded the established national organizations on more than one occasion. Whether manifested by ecotage or philosophical quarterly, by seminar or demonstration, the radicalism of the eighties played a real role in raising the issues, and the stakes, of American environmentalism.

*   *   *

By the end of the Reagan presidency in 1989, and in defiance of the Reagan Reaction, the environmental movement in all its guises was stronger than it had ever been. It had certainly grown polymorphously, in the capital and at the grass roots, its various streams themselves holding many different types of organizations with a great diversity of resources and multiplicity of aims—this is a movement that went from the conservative Izaak Walton League and senior citizens' birding clubs to the radical Sea Shepherd and monkey wrench gangs—but it was an undoubted presence on the national scene. It was there in the professional power of multimillion-dollar national offices operating in the highest corridors of decision making, there in the popular plethora of NIMBY-energized groups now in almost every hamlet of the nation, there in the broad caldron of creative and influential radical thought and direct action, and probably in a myriad of other groups and influences that fit into the interstices and defy categorization. It was at any rate strong enough to withstand the backlash of a popular and persuasive President—as Stewart Udall noted, "It is clear that on environmental issues Ronald Reagan rowed against the American mainstream for eight years"—and to emerge with values and aims and energies intact.

5

ENDANGERED EARTH

1988–92

I N the summer of 1988, the earth sent humankind a message. In the United States, a persistent drought gripped the West and much of the South, killing off livestock by the thousands and drying up grainfields so thoroughly that a third of the crop was lost. An unprecedented heat wave, with record-breaking 100-degree temperatures across much of the country, seemed to foretell of the global warming predicted by scientists as a result of carbon gases surfeiting the atmosphere and creating the "greenhouse effect" of an ever-warmer climate. People seeking relief at the beaches found them fouled with garbage and raw sewage along the Atlantic and Gulf coasts, and those who chose the western national parklands found many of them in flames and covered with smoke, victims of the worst forest fires in a generation. Waterways throughout the Midwest experienced their lowest flows in years, and the level of the Mississippi was down so far in the summer that navigation was curtailed over long stretches. Newspapers told of increasing acid rain, massive radioactive wastes at the nation's weapons plants, holes in the earth's protective ozone layer, bizarre bioengineering experiments, city landfills used up and closed down across the country, endangered species disappearing at the rate of two an hour, and the worldwide journey of the freighter *Pelicano*, which had spent two years hunting for a port that would accept its cargo of 14,000 tons of toxic incinerator ash.

Elsewhere, the world was equally out of whack. Hurricanes more powerful than usual lashed through the Caribbean, floods killed millions in Bangladesh, and droughts crippled harvests in China and the Soviet Union. Pollution was so severe that beaches on the Mediterranean and the English Channel were closed for long periods. The Amazon rainforest, already being destroyed at the rate of a football field a second, was burned with unprecedented profligacy by farmers and developers in fires that sent dark smoke over several million acres and as far north as the southern United States. And surveys of population growth indicated that the numbers in Africa and Asia were continuing to increase by 4–5 percent a year, thus threatening to double already deprived populations in less than twenty years, far outstripping any known agricultural resources.

Time, which decided that the summer's crisis so dominated the news that it awarded its usual man-of-the-year issue to a construct it called "Planet of the Year," wrote darkly but not inappropriately: "This year the earth spoke, like God warning Noah of the deluge. Its message was loud and clear, and suddenly people began to listen, to ponder what portents the message held."

And so, at least on a superficial level, it seemed. A foot-dragging Administration in Washington gave way to a new President who had actually included environmentalism among his platform planks—for the first time on the Republican side —and dared to proclaim himself, albeit for votes alone, "the environmental candidate." The environmental establishment ratcheted itself up to a new pitch, increasing memberships and budgets and refining strategies, and went all out to make the twentieth anniversary of Earth Day 1970 its offensive catapult for the nineties. And both within the United States and without, the world gave indications beyond any perceived before that it had at least heard the message and would ponder its foreboding import.

If, after all this, the environmental movement regarded the future with a great deal more seriousness than perhaps ever before, and if it was inclined to be aware of how much remained undone rather than how much it had accomplished, that was no doubt a measure of the true dimensions of the ecological crisis and a deep understanding of the implications of an endangered earth.

In the election campaign that followed the scary summer of 1988, it was, surprisingly, George Bush who talked more about the environment than his Democratic opponent, Michael Dukakis, and for the first time the issue played a prominent part in campaign rhetoric and advertising, apparently to Bush's advantage. A Bush TV commercial attacking Dukakis for the foul pollution of Boston Harbor, depicted as a pool of sewage, was credited by some pollsters as decisive in damaging the Massachusetts governor's image nationally.

Bush's term in office, however, was hardly an environmentalist's dream. The one notable appointment was William K. Reilly, executive director of the powerful Conservation Foundation! World Wildlife Fund, as head of the EPA; Reilly, though a moderate noted for working with (and accepting donations from) large corporations, including chemical and oil companies, was also known to be critical of existing government programs, including the stalled Superfund, and in favor of a "massive restructuring" of Washington's environmental operations. He was, however, effectively overbalanced by the appointments of Manuel Lujan, an anti-environmental congressman from New Mexico, as Secretary of the Interior, Richard Darman, a conservative numbers-cruncher, as head of the OMB, and (until early 1992) John Sununu, an unregenerate foe of environmentalism, as the important White House chief of staff. Not to mention the Chief Executive himself, a Texas oil millionaire and oil industry champion—and ardent hunter and fisher—who as Vice President had run up a solid record of intervention on behalf of corporations seeking exemption from EPA regulations on everything from hazardous-waste disposal to unleaded gasoline.

Though environmentalists personally presented the new President with a "Blueprint for the Environment" shortly before he took office—along with a warning from former EPA administrator Russell Train that "the problems hitting us now are hellishly more complex and difficult"—the Bush Administration managed generally to avoid taking action for four years on all but the most pressing issues. In doing so it raised to a new high the art of administrative neglect—having its appointees and bureaucrats, often in the secrecy of the White House, dilute, rewrite, or simply ignore regulations established by Congress and the EPA—with such effect that *The New Yorker* was moved to complain that it was a pattern "consistent enough to suggest that the Administration has made a conscious decision to operate this way, so as to be widely perceived as pro-environment without alienating any of its corporate supporters."

Bush was given credit for breaking the 1990 Clean Air Act amendments out of a Congress where action had been stalled in a legislative logjam for more than a decade, though the White House succeeded in substantially weakening the provisions before passage and having its Council on Competitiveness weaken and bypass many of them afterward. He reluctantly agreed to accelerating the end of CFC production from 2000 to 1996 or sooner, but only after presented with uncontestable proof that the ozone layer in the Northern Hemisphere—and over his summer house in Maine—was growing dangerously thin. Bush also got occasional headlines for decisions on a number of painless and low-cost issues: blocking a dam on the Colorado River in 1989, reducing air pollution in the Grand

Canyon, granting a ten-year moratorium on most offshore drilling, voting for a UN ban on ocean drift-net fishing, and lowering thresholds on lead poisoning, all in 1991, and extending the Endangered Species list by 50 percent in 1992.

The Bush Administration also acted forthrightly in continuing the multibillion-dollar program of cleaning up contamination at military bases and weapons plants where billions of gallons of radioactive wastes, chemical by-products, and toxic liquids were buried from 1945 on. But this program, which will eventually cost at least $400 billion—the largest and most expensive engineering project ever undertaken, four times that of the space program—was started before Bush came to office and has been pushed by the military itself as a result of community pressure and increasing complications at the more than 18,000 sites that have produced what *The New York Times* has called "environmental contamination on a scale almost unimaginable." (Worldwide, according to the Science for Peace Institute at the University of Toronto, "10 to 30 percent of all global environmental degradation can be attributed to military activities.")

On most other fronts the Bush Administration won no environmental plaudits, and its performance grew worse as the term went on. On the international front, the Bush Administration refused to sign a treaty on carbon dioxide emissions in the atmosphere, accepted by all the other nations of the world, until it was watered down to meaninglessness, and at the 1992 Rio "earth summit" it was alone in opposing and refusing to sign international accords on preserving endangered species and protecting forests and wetlands. At home, Bush made development of the Arctic National Wildlife Refuge for oil drilling the centerpiece of his energy policy (the measure was shelved by Congress in 1991), while taking no action on reducing energy consumption or increasing automobile efficiency. A series of Administration policies enacted in 1991 and 1992 opened up vast new areas of land, including wilderness, wetlands, and old-growth forests, to coal, timber, and oil interests in what *The New York Times* described as "the strongest effort to reduce environmental restriction" since the early days of the Reagan Administration. Typical of such moves was one three-month period in 1992 when the Interior Department sided with the coal industry against enforcement of strip-mining laws, the Forest Service pressed to open up protected national forests to clear-cutting despite acknowledged habitat destruction, and the EPA came out with proposals to weaken hazardous-waste restrictions. It was a record, according to liberal *New York Times* columnist Anthony Lewis, that entitled Bush to be known henceforth as "the pillage President."

But perhaps the most serious environmental act of the Bush Administration was the war in the Persian Gulf in January–February 1991. Nothing was done to anticipate or prevent environmental damage, though it was surely predictable (and indeed predicted by many environmentalists), and so when some 300 million gallons of crude oil were poured into Gulf waters in the worst oil disaster to date, more than 700 oil wells were set on fire to spew toxic chemicals (including 50,000 tons of sulfur dioxide and 25 tons of carbon dioxide a day) into the air all over the region, and desert ecosystems were obliterated by bombs and machinery, nothing could be done to halt or mitigate its calamitous effects. Surprisingly, most mainstream environmental organizations raised no particular opposition to the war, a task left to representatives from Earth First!, Earth Island Institute, Friends of the Earth, Greenpeace, Environmental Action, U.S. Greens, and a few others in a statement issued that January; the staff of Greenpeace went further, issuing a sharp attack on American and allied intervention that concluded, "Ecologically, the war in the Persian Gulf is a consequence of a fundamentally destructive way of life, centered on our addiction to oil."

George Bush summed up his term with the boast that his was

a balanced policy: "Our Administration," he said, "has crafted a new common-sense approach to environmental issues, one that honors our love of the environment and our commitment to growth." But the evidence was clear that when the two came into conflict, as is inevitable with an economy built upon resource depletion, the latter almost always took priority. The true sentiments of the Bush regime were perhaps best expressed by OMB head Darman in May 1990: "Americans did not fight and win the wars of the twentieth century to make the world safe for green vegetables."

Surveying the impact of environmentalism in 1988, Denis Hayes, the 1970 Earth Day coordinator, noted that "in terms of sustainability, the movement has exceeded our expectations. It's managed to avoid the American tendency to come and go like Hula-Hoops, Davy Crockett hats, and punk-rock haircuts"— and, he might have added, like a similar trend, the peace movement, whose leading organizations, SANE and Nuclear Freeze, were already floundering despite the challenge of a massive Reagan arms buildup.

By almost any measure, the movement, both mainstream and radical, was alive and well and ready to take advantage of the fears unleashed by the summer of 1988, and to grow in the next four years. As one Washington environmentalist reported, "Since 1988 environmental issues have really advanced to center stage, and now we know how to be the directors and the producers—and even the lighting designers."

To begin with, both the number and variety of organizations, Topsy-like, just kept growing. In addition to the quite uncountable number of local and regional groups—as many as 12,000, according to one 1990 estimate (some 1,300 of them, for example, affiliated with just one coalition, the National Toxics Campaign, in 1990)—there were at least 325 organizations of national standing, according to the 1991 *Encyclopedia of Associ-*

ations; of those, the most active and important, as listed in two environmental directories that came out in 1991, numbered between 100 and 150. The assortment went from the Animal Protection League to Zero Population Growth and in between included (counting just some begun in the latter half of the eighties) Ted Turner's Better World Society, Center for Whale Research, Conservation International, Earth Communications Office, Kids for a Clean Environment, International Marinelife Alliance, Pacific Wildlife Project, and Rainforest Alliance. Among the most important groups, newly formed or newly active, were:

- Student Environmental Action Coalition, which began as an idea on the University of North Carolina campus in 1988, blossomed by 1992 to embrace 33,000 students on 1,500 campuses and at 750 high schools, its organizers said. Through a regular newsletter, a network of campus organizers, a series of coast-to-coast "working groups," and three national conferences, SEAC revived campus politics in a way reminiscent of the 1960s and set itself the task of "radicalizing" the "rich" mainstream environmental groups to pay more attention to social justice issues and urban environments.

- Animal Legal Defense Fund, which grew out of Attorneys for Animal Rights in 1984, became active in the late eighties, using courts to stop deer, bear, and lion hunts, bringing suits against cosmetic and medical animal research, and encouraging student protest against animal dissection in the classroom. Like other animal rights groups in a movement now newly energized—Citizens to End Animal Suffering and Exploitation, Progressive Animal Welfare Society, People for the Ethical Treatment of Animals—it also gained attention for protests against keeping dolphins and whales in captivity and using them in scientific and military experiments.

- Rainforest Action Network, founded by Earth First!er Randy

Hayes in 1985, made its mark with a boycott against Burger King in 1987 to protest the use of beef from countries that allow clear-cutting of forests to make room for cattle ranches. That campaign not only forced Burger King (and other fast-food chains) to change its buying policies; it brought the issue of rainforest destruction forcefully to the attention of the public, where it has remained one of the two or three most popular causes. By 1992 there were more than 250 organizations worldwide working on rainforest protection and RAN itself had grown to more than 30,000 members, with more than 150 "action groups" nationwide set up to run local direct-action campaigns, pressure officials and lawmakers, and alert the network (by mail and computer) to companies guilty of forest destruction.

- Society for Ecological Restoration, begun in 1988 after two decades of both practical and theoretical work in revivifying ecosystems, had 1,700 members by 1991 and a wide national influence through its quarterly *Restoration and Management Notes*. The restoration movement had a growing following among both academics and hands-on spade-and-seeders (and such organizations as the North American Bioregional Congress and the Nature Conservancy), and according to the *Whole Earth Review* in the spring of 1990 *"a lot* of work" was being done on coastal, wetlands, prairie, forest, and stream restoration from the Mattole Valley in California to the Hudson River estuary in New York.

Though the figures are not always reliable, the membership in environmental organizations of all kinds was clearly on the upswing during this period. Calculating from the 1991 *Resource Guide to Environmental Organizations*, total membership in the larger national organizations could be estimated at around 20 million, with perhaps a 30 percent overlap, or more than 14 million individuals in all, about one in every seven adults in the land. Figures for the majors alone were impressive, showing steady and sometimes spectacular increases; as of 1991 the total for the Group of Ten stood at nearly 8 million:

Audubon Society	600,000
Defenders of Wildlife	80,000
Environmental Defense Fund	150,000
Friends of the Earth (*incorporating the Environmental Policy Institute*)	40,000
Izaak Walton League	50,000
National Parks and Conservation Association	100,000
National Wildlife Federation	5,600,000
Natural Resources Defense Council	170,000
Sierra Club	650,000
Wilderness Society	350,000
	7,790,000

That doesn't count some of the other majors, which—like Greenpeace at 1.5 million (and another 3.5 million overseas) and World Wildlife Fund at 900,000—were quite sizable.

Growth of this magnitude, of course, reflected a wider fact: the considerable popularity of environmental protection among the general public, particularly pronounced when any large-scale environmental crisis occurred. Survey after survey indicated that Americans took environmentalism seriously: a 1990 CBS poll, for example, found that 74 percent of those questioned said that protecting the environment was so important that no standards could be set too high (up from 45 percent in 1981), and nearly half of them proclaimed a "strong identification" with environmentalism; a 1990 Gallup poll found that 76 percent of Americans called themselves environmentalists and half contributed to environmental organizations. When catastrophes struck—as they did with alarming regularity—much of this support was translated into membership and money. After the

Exxon Valdez ran aground in Alaska in March 1989, for example, spilling 11 million gallons of crude into the once-pristine Prince William Sound and forcing a $3.5 billion cleanup operation, almost every environmental group reported a rash of new members; not only that, but the marketing and membership consultants lost no time in following up with direct-mail appeals, as at NRDC, which sent out 500,000 pieces of mail in the weeks following.*

Concomitantly, the coffers of the leading organizations also grew apace. It was estimated that the Group of Ten had budgets amounting to something around $250 million in 1990, of which these are representative:

Audubon	$42 million
EDF	17 million
NWF	87 million
NRDC	16 million
Sierra	23 million
Wilderness	18 million

and that the other national organizations had budgets of a similar amount, perhaps $550–$600 million in all.† This was not the kind of money power that corporations could muster for environmental issues—industry anti-environmental lobbying and PR groups had budgets that easily matched this, and corporations could spend an estimated $25 million a year in green elections alone—but it was significant in terms of public-

---

* Sometimes it was not even necessary to wait for the catastrophe. In 1989, for example, the NRDC hired a publicity firm to draw attention to its study showing that the chemical Alar, used by many apple growers to enhance fruit shelf life, was carcinogenic, and the resultant exposure of it on the *60 Minutes* TV program caused a storm of controversy so fierce that apple sales plummeted and the growers stopped using it. NRDC membership increased dramatically.

† *USA Today* (September 5, 1989) estimated a total of $420 million for the budgets of the twenty largest environmental organizations.

policy movements in the United States and it allowed environmentalism to cut a much larger swath than ever before. It was,
moreover, embedded in a much larger arena of environmental
giving of all kinds—foundation grants to individuals and causes,
for example, individual sponsorship of particular projects, be-
quests and trust funds, workplace donation campaigns, local
fund-raising—that was estimated to amount to as much as $2.5
billion in 1991, up from $1.3 billion in 1987.

The high point of this new sense of power was probably Earth
Day 1990, a twentieth anniversary celebration of the first Earth
Day, only this time made into a powerful, expensive, widespread,
and media-suffused event. Denis Hayes was again coordinator
of the main efforts in the United States, but this time he had
eighteen offices across the country and staffs that numbered in
the hundreds, high-powered boards of advisers and even more
high-powered corporate donors (total budget: $3 million), and
an agenda of some 3,000 events coast to coast including an
array of TV specials with big-name stars. The day itself—April
22, 1990—was celebrated by what was estimated to be more
than 100 million people in 140 countries, including perhaps 25
million in the United States. The national feeling here ran high,
with support from politicians of virtually every stripe and evident
outpourings of enthusiasm from classroom to boardroom, but
there was no direct legislative impact as there had been in 1970,
not even any specific legislative wish list. The primary effect
after all the hoopla had died down seemed to be commercial:
not so much the array of merchandise got up for the day itself
(including no fewer than two dozen books) as the decision by a
great many companies to "go green," offering safer or environmentally friendly products, using recycled (or in some cases
"recyclable") materials in packaging and promotion, and in
general putting out the image of environmental responsibility
and good citizenship, sometimes genuinely, sometimes not; this
led to the creation of two firms designed to test products for

"green consumer" labels, one of them headed by Denis Hayes.

But the impact of Earth Day 1990 was not so much tangible and practical as psychic and moral, a declaration by the body politic of its concern and its passion. *The New York Times* was right to say after the event: "The environment is not just another issue. It has become a modern secular religion."

It was in this context that the mainstream environmental organizations pursued their strategies during the Bush years, and with some measure of success, too. Peter Borrelli of NRDC's *Amicus* quarterly pointed out that the national offices "have become strategically deft at carrying out the agenda: lobbying Congress one day, going to court the next, and meeting with labor and religious leaders the next." Newly enlarged staffs, with newly professional tools, could operate in the national arena with perceptible effect and sometimes, as a Sierra Club lobbyist put it, "with real political clout."

In truth, though, the record was decidedly mixed. The Bush years were, all in all, convincing evidence of the environmental paradox: power and progress, popularity and performance, but still the goals were elusive, the problems formidable, the crises unabating. The environmental sword, however sharp, seemed always to confront the ecocidal hydra.

Take, for starters, the most touted and probably most controversial strategy of the majors, called "third-wave environmentalism" by its adherents—especially EDF, Nature Conservancy, World Resources Institute, and World Wildlife Fund—and appeasement by its critics. It was a systematic attempt to work with the movement's traditional enemies, corporate polluters and extractors, to achieve by cooperation and reason what couldn't be done by confrontation and regulation—or, as the head of the National Wildlife Federation put it, "enlightened and responsible dialogue with corporate leaders." Among its several successes were the negotiations by which Earth Island

and other groups persuaded large West Coast tuna fleets and packers to give up fishing practices that endangered dolphins and to market "dolphin-free" tuna; and the agreement between the Audubon Society and a variety of large consumer-product firms in which the companies accepted model legislation reducing heavy metals in their packaging within two years. On the other hand, discussions between environmentalists and businesses in California to spell out in detail the proposed effects of the 1990 "Big Green" referendum ended up with a forty-page bill of such complexity that no one could understand it and it failed at the polls; and meetings begun by the National Wildlife Federation in 1989 between the EPA's William Reilly and North Carolina toxic-waste executives led to new moves *against* a North Carolina law that environmentalists had passed to keep the notoriously polluting waste facilities out.

The much-heralded agreement in 1991 by which the Environmental Defense Fund's Fred Krupp persuaded McDonald's to abandon its ubiquitous polystyrene packaging at its fast-food outlets was perhaps typical. A five-year campaign by various grass-roots groups to force the fast-food chain to switch had been stonily resisted, but after Krupp became friendly with McDonald's president, Ed Rensi, and assigned a three-person EDF task force to work with his executives to find alternatives to the unbiodegradable polystyrene, McDonald's finally decided to start using paper wrapping throughout its empire. Krupp was jubilant: "It proves that McDonald's recognizes the future is green." But one Washington lobbyist was less enthusiastic: "You could hardly call it a victory when they're just going to spew out in the countryside the same *amount* of packaging, only now they're going to cut down whole forests to supply it, and one of the papers isn't even biodegradable. We should be about real reductions in corporate packaging, not just different kinds." If not a victory, though, not a failure—perhaps best seen as an ambivalent success.

Or take, similarly, the various initiatives that made up the new international strategy of these years, inspired both by a new perception that so many environmental problems were global—acid rain, ozone depletion, atmospheric warming, whale overfishing, Antarctic mining, deforestation—and by a new attention to them by world leaders, most especially the Soviet Union's Mikhail Gorbachev.* Both old and new groups in the United States responded: Greenpeace and the National Wildlife Federation, for example, fought for laws to ban the international trade in toxic chemicals, Earth Island went to the defense of Central American forests, EDF pushed for an international ban on dumping plastic waste at sea, Greenpeace took an activist position on the International Whaling Commission, newer groups like the Global Greenhouse Network and the Better World Society drew public attention to global warming, the Campaign to End Hunger and LIFE (Love Is Feeding Everyone) worked on malnutrition and starvation worldwide, and Conservation International and World Resources Institute studied and initiated environment-protective "debt-for-nature swaps." This last program, one of the most innovative of this era, was designed to find private funds to cancel part of a country's foreign debt to American banks (or persuade those banks to forgo collection) in exchange for the country's promise to ensure permanent protection of a threatened ecosystem; begun in 1987 with an expansion of the Beni Biosphere Reserve in Bolivia to protect 4 million acres of tropical forest when Conservation International paid Citicorp $100,000 for an uncollectable $650,000 loan, it had encompassed more than $100 million in bank loans and effected twenty swaps by early 1992.

There's no doubt that concerted work and publicity by these various activists—and the continuing eruption of global crises

* As of 1989 there were environmental agencies in more than 140 national governments around the world, and more than 250 international environmental agreements signed.

—helped create the climate that forced environmental issues onto the global agenda.* It was largely responsible for the dramatic Montreal agreement in January 1989 in which most of the world's nations pledged to reduce chlorofluorocarbon production in five years and phase it out where at all possible in ten; and for the international agreement in Madrid in October 1991 establishing the Antarctic as a "world park" in which mineral and oil exploration would be banned for at least fifty years. But aside from those achievements there was precious little to point to in the way of serious action on the international front.

The industrial-nations ("Group of Seven") summit in 1990 was forced to devote nearly half its time to environmental issues and resolutely called for "decisive action" to be taken, but there was no follow-up at all and in the 1991 summit a scant thirteen minutes was devoted to sweeping the issues under the rug for another year. The UN's Brundtland Report (*Our Common Future*), released in 1987 after four years of studies, proved to be a near-empty document (except for the tables and boxes spelling out the crises in detail), quite happy with pushing the idea of "sustainable development" within a context of continuing population growth, present multinational corporate power, existing monetary systems, increased exploitation of nonhuman life forms, and increased dependence on modern technologies, including pesticides and "agrochemicals." The World Bank, despite new lip service to "ecologically sound patterns" in 1990, continued to spend nearly $20 billion a year on large-scale projects that destroyed ecosystems and uprooted some 1.5 million people worldwide, acting, in the words of the *New Internationalist*, like "an ecological Frankenstein armed with a

---

* Plus, as well, the unintended wave of support that followed such misconceived acts of resistance as the sabotage of the Greenpeace ship *Rainbow Warrior* and the death of one of its crewmen in Auckland Harbor in July 1985, planned and carried out by the French government, which wanted to prevent the ship's monitoring of a French nuclear test in the Pacific.

chainsaw" in a "full-fledged assault on the remaining tropical and temperate forests" of the Third World. And the worldwide Green movement, though it showed some strength in the European parliamentary elections in 1989, seemed effectively stalled, with the German Greens badly split and harshly rejected in the 1991 reunification elections, the American Greens finally unified but not attracting much political attention, and the European Greens nowhere with more than 10 percent of the vote and nowhere playing the role that they had envisioned a decade earlier.

As for the United Nations Environment Programme, now in its second decade and still largely ineffective, it decided to place all its emphasis on the "Earth Summit" in June 1992 in Rio de Janeiro, a replay of the UN's Stockholm Conference but with an even greater number of nations represented (143), at least two-thirds of them by heads of government. In a lengthy series of preliminary meetings, an international consensus was established on treaties for reducing carbon emissions that contribute to global warming, for protecting species threatened by development and pollution, and for preserving forests fast falling before timber and cattle industries, all of which were to get a final global stamp at the summit. Unfortunately, however, the United States—the world's largest economic power and the world's largest polluter, consumer of energy, and generator of wastes—refused to go along with the rest of the world, effectively gutting the global warming treaty and flatly rejecting the other two, then refusing any commitment to a scheduled "North-South" fund through which industrial nations would help developing ones over the hump to "sustainable" economies. In the event, although the Rio meeting attracted representatives of some 6,000 grass-roots organizations from around the world and provided the forum for a World Conference of Indigenous Peoples and the First Planetary Green Meeting, the official side of the summit was largely platitude and rhetoric.

What Rio did reveal, though, was the underlying agenda for

the global environment that the United States and many of its industrial allies had been crafting for some years. The Rio nations agreed to create a new Global Environmental Facility, effectively controlled by the industrial nations and administered by the World Bank, that would monitor future environmental threats and determine appropriate responses, governed by the general principles of enhancing free trade and enlarging world markets. Such a body would essentially carry out the strategy spelled out by the World Bank's *World Development Report*, issued just before the summit, according to which future ecological harmony was to be established by three mechanisms: population control, achieved by putting more Third World women into the work force; fine-tuning market responsiveness, to reward careful resource exploitation and reasonable "pollution permit" trading; and increased private property in Eastern Europe and the Third World so as to encourage individuals to develop land in their own self-interest. Combined with the industrial nations' General Agreement on Tariffs and Trade, attempting to regularize the work of global corporations throughout the world, the new Global Environmental Facility was a sure step toward the kind of environmental regulatory climate that would serve most the interests of the developed industrial powers.

Lastly, let us take the three means by which mainstream environmentalism continued to direct its efforts in the Bush years: legal, electoral, and legislative. Again, the record is one of ambivalent success: not by any means negligible, but well below what most environmentalists had hoped for and most Americans had declared themselves in favor of, the shadows of ecological crisis looming over even the most brilliant victories.

Though the legal approach was shunned by some third-wavers as overly confrontational—the EDF said that "rather than go to court we lobby, write reports, court the media"—it was still the favored tactic of groups like the NRDC and the

Sierra Club Legal Defense Fund as well as an increasing number of private law firms. "Law will continue to be one of the environmental movement's most potent tools," said Tom Turner of the Sierra LDF at the end of the eighties. "It works, and it rewards its practitioners." Certainly the number of cases continued to grow—the *Environmental Law Reporter* published more than 4,000 federal court decisions between 1971 and 1988, most of them in the latter decade—and a number of blatant offenders of environmental laws, particularly the furtive "midnight dumpers" of toxic wastes, were brought to the bar and found guilty, a total of more than 300 convictions involving nearly 100 corporations in the 1980s. Certainly also the number of environmental damage suits increased, at federal and state levels alike, with asbestos lawsuits absolutely jamming the dockets (more than 1,140 suits a month filed in 1990 and a backlog of more than 30,000 cases in the courts waiting to be tried).

And yet not many in the legal process claimed that all this was anything more than a kind of rearguard, largely defensive and often tardy, action. Corporations had batteries of lawyers to fight environmental cases, expense be damned, and they became not only increasingly resistant but increasingly combative as time went on; one favorite tactic from the mid-eighties on was called the SLAPP—Strategic Lawsuit Against Public Participation—filed in the hundreds by developers and polluters against environmentalists seeking to impede their businesses, embroiling individuals and community groups in costly (average: $9 million) and lengthy (average: 36 months) lawsuits regardless of outcome. Moreover, government agencies themselves could be as difficult as industrial opponents, refusing to enforce regulations or ignoring statutes, and in that quicksand many a lawyer spent many frustrating months and energies. One veteran environmental lawyer, Rick Sutherland, president of the Sierra LDF, offered this assessment in 1991: "My primary emotion when recalling the past twenty years of environmental law is one of profound disappointment . . . due to the continuing failure of federal agencies and officials to do a better job of implementing and enforcing our environmental laws." In fact, he said, the files of legal-defense organizations were filled with suits not against illegal corporations but against agencies of the federal government, particularly the Forest Service, the Fish and Wildlife Service, and—astonishingly—the EPA itself. And if that wasn't enough of a problem, the Supreme Court at the top of the legal system had shown "increasing hostility" toward environmental issues, deciding every single one of the EPA suits brought before it against enforcement of the laws and in all its other cases proving "unsympathetic to environmental concerns."*

The electoral approach, spearheaded still by the League of Conservation Voters and various environmental PACs but with more and more state and local organizations involved, was not without its impact, though both the 1988 and 1990 congressional elections were setbacks. In the 1988 presidential election most of the environmental groups were in the Dukakis camp (LCV, for example, gave Dukakis a "B" and Bush a "D") and were embarrassed by his defeat, but 60 of the 76 congressmen backed by LCV were elected and 17 of 24 local initiatives were passed. Much more was expected in the 1990 campaigns—when, *The New York Times* declared, "candidates for one public office after another [are] proclaiming themselves environmentalists"—and LCV doubled the money it spent on candidates (still only $300,000, paltry compared with industry PACs) and expended more than twice as much to produce and distribute radio and

---

* The following year the Court went on to overturn one of the basic strategies of legal environmentalism, ruling that ordinary citizens have no legal standing to challenge government rulings on the Endangered Species Act unless they can show direct and imminent personal injury, not merely general concern for habitat and life on earth.

TV ads; for the first time major environmental initiatives were put on the ballot in California ("Big Green," the most comprehensive proposal ever offered in the nation), New York, Oregon, Massachusetts, and Missouri. In the event, though, 1990 was a disappointment: all the initiatives lost—Big Green by a crushing 64 to 36 percent—and of LCV's 133 candidates only 84 were elected.

Various sober lessons were drawn after the 1990 campaign about the limits of electoralism. One obvious conclusion was that environmentalists could never outspend corporations, an increasingly important factor as campaigns became more and more expensive; after raising an impressive $4.7 million in California, for example, supporters of Big Green were overwhelmed by the $17 million put up by oil, timber, and agribusiness interests. Another lesson was that voters were not much inclined to endorse projects in which government agencies were charged with spending lots of money, no matter how environmentally useful—a rejection, as *Greenpeace* editor Andre Carothers put it, "of the top-down, expert-laden, publicly financed programs that the average voter has seen come and go, with little effect."

Not much more hope was generated by the 1992 Presidential elections, either, despite the fact that a leading environmental politician, Senator Albert Gore of Tennessee, was on the ticket with the winning Democrat, Arkansas Governor Bill Clinton. Environmental issues played almost no part in the campaign itself—exit polls showed that fewer than 10 percent of the voters said the subject was important in their decision—and no major referenda were on the statewide ballots. The principal environmental organizations, hurting already from a fall-off in income as a result of the long recession, played only modest roles during the campaign, and though the League of Conservation Voters spent a record $600,000 in support of environmental candidates, only 108 of the 186 people it endorsed for federal office were

elected, a 58 percent showing, down from 63 percent in 1990.*
In the end it was the Gore victory that seemed to hold out any
hope for an environmental presence in the succeeding Administration—he was, among other things, author of *Earth in
the Balance*, a best-selling book on ecological peril and its presumed remedies—but with a new President marked by a shabby
environmental record in his home state and preoccupied with
economic recovery on the national agenda, the prospects were
not bright.

The electoral strategy, too, had been hit by the harsh reali-
zation that the election of green candidates—or even olive or
khaki—did not always mean the passage of green legislation,
an acknowledgment of the limitations of the legislative strategy.
Although environmental groups continued to lobby mightily—
led by the Sierra Club, NWF, and especially NRDC, whose
influence *The Wall Street Journal* called "profound"—and though
they were acknowledged by former Interior Secretary Stewart L.
Udall to "exert a powerful influence on lawmaking in Washington," the fact was that not much got passed, too little was
blocked, and a great many hopes were dashed.

Legislation on wilderness and timber areas, issues that revived
considerably in importance around this time, provided a case
in point. Bills that would ban clear-cutting in the national forests
or on federal lands and protect large areas of unspoiled ecosystems were put in the hopper but, despite environmentalists'
claim to having a "power base" and "swing votes" in the House,
none of them was even seriously considered in committee; in
fact, the Sierra Club even went so far as to discourage action
on some of them, hoping to keep its reputation for "reason-
ableness" intact. Most of the majors' efforts, in fact, were devoted
not to passage but prevention, and there the only important

---

* Candidates running on avowedly Green tickets took part in 85 races, mostly
in California and Hawaii, but only 13 of them were elected (raising the number
of Green officeholders to a meager 58), largely to school and water boards.

victory was endorsing a threat of a filibuster that eventually ended action on Bush's plan to open the Arctic National Wildlife Refuge to oil development. In between all that, when appropriation bills came up for the EPA and the Interior Department it was all these high-powered outfits could do to minimize cost-cutting damage, even then being forced into compromises (the Sierra Club agreeing to *doubled* timber sales in the 1990 Interior appropriation bill) that often outraged activists in the field. As one Montana ecologist put it, more in sorrow than in anger, the majors "have strayed from their grass-roots origins" and "cues are taken from the corporate world," largely because of the heavy influence of "tired bureaucrats who do not spend nearly enough time in the wilderness." A colleague of his, after losing a fight on six million acres of virgin forest, was more bitter: "They talk like Rambo in their fund-raising letters," said Tim Hermach of the Native Forest Council in Oregon, "but the big national groups are wimps when it comes to dealing with the bad guys."

The conclusion to be drawn about the effect of environmentalism by the end of the Bush Administration, then, was by no means as sanguine as it had seemed in the heady days of the early seventies or in the great flush of attention around Earth Day 1990. The new third-wave approaches as well as the older legislative and electoral strategies clearly had shown their limits, and disappointments and defeats had to be counted along with the important gains. If it was true that the movement now was entrenched, efficient, and often effective, if it could claim credit for the passage of hundreds of new laws to protect the landscape and its denizens, if it was responsible for saving millions of acres of lands from rapacious development and hundreds of thousands of wild creatures from extinction, if it could point to rivers cleaned and toxics banned and polluters shut down—it was also true that more than half the population of America lived in counties in violation of Clean Air statutes, 5 tons of carbon per person per year were pouring into the atmosphere, at least 170,000 lakes and millions of acres of forests in the United States and Canada were acidified, 90 percent of the garbage produced went unrecycled, less than 5 percent of the worst toxic-waste sites in the nation had been treated, topsoil was being washed away at the rate of 3 billion tons a year, water was depleted or polluted at the rate of 10 billion gallons a year, and the alteration of minds and habits necessary for the salvation and health of the world had only barely begun. Even Denis Hayes, normally optimistic in the atmosphere of Earth Day, admitted that "by any number of criteria that you can apply to the sustainability of the planet, we are in vastly worse shape than we were in 1970, despite twenty years of effort."

And there is the paradox. The earth had indeed sent a warning in the summer of 1988, as it often had through these years, and it was a message that by and large the citizens of the United States (and much of the world beyond) had taken seriously and to heart. In response, an enormous amount of time and effort and money had been spent by an enormous number of people, disinterested people of goodwill and deep commitment, and there were indeed visible effects as a result. The earth, however—and this sad truth lay at the heart of the matter—was still in danger, and seriously so, and the task of the environmental movement seemed to be only just begun.

6

PROSPECTS

"WE'RE not where we *want* to be," Martin Luther King, Jr., used to tell his civil rights audiences in the 1960s. "And we're not where we're *going* to be. But we sure are a long way from where we *were*!"

So much done, so much left undone: such is the way with movements for social change. Such is the way with the environmental movement, only it is a movement concerned not just with human rights but with human survival—and the survival of all other species and the ecosystems that give them life. Thirty years after its effective inauguration with Rachel Carson's *Silent Spring*, after its transformation of the dedicated but limited conservation movement into a vital and widespread social and political phenomenon, the environmental movement could point to all kinds of significant achievements, even as it had to acknowledge all kinds of disheartening failures. And if it was true that the victories outnumbered the defeats, and seemed in retrospect to be more substantial and spectacular, still there was always the sober wisdom of environmentalist José Lutzenberger to consider: "In the environmental movement, our defeats are always final, our victories always provisional. What you save today can still be destroyed tomorrow, don't you see?"

Assessing environmentalism after its first three decades—and considering where it might be heading in the foreseeable future—is a task of great complexity, not only because the movement is so extensive and protean but because there are the many contradictions inherent in the experience of inter-twined victories and defeats, steps back and steps forward, campaigns successful and strategies not. But there are some measures that can be taken, some signposts that can indicate how much of the road has been traveled and how much is still to come.

To begin with, it is only right to restate the considerable achievements. Within the space of a single generation, environmentalism has become embedded in American life, in law and custom, text and image, classroom and workplace, practice and consciousness, in such a way that suggests that, far from eventually fading away, it will continue to grow, in numbers and impact alike, at least for some time to come. It is embedded in national legislative and administrative institutions—most strikingly the Environmental Protection Act of 1969, the Clean Air and Clean Water acts, and the Endangered Species Act of 1974, all of which have been responsible for regulations now estimated to cost the society $125 billion a year—and in fundamental judicial decisions giving the environmental viewpoint legal standing in countless arenas of public life. It is embedded in the acts of individuals, from schoolchildren to CEOs, in the functions of communities, from villages to block organizations, in the performance of governments, from city water departments to national administrations. It is embedded in political life, where it is both an electoral and nonelectoral force, in economic life, where it now affects billions of dollars' worth of decisions each year, in cultural life, where it is a daily part of journalism, publishing, education, and the arts, and in social life, where it affects habits, vacations, clothing, food, travel, even friendships.

Along with this embedding have come certain home truths,

by now well-nigh inescapable. Citizens of the United States (and most other nations) are understood to be vitally concerned about the quality of their physical world, ranging all the way from the importance of clean water, air, and food to the sanctity of wilderness and its diverse species. Governments at all levels, from local to global, are seen as having a vital role in the protection and preservation of the environment, taken in the very broadest sense, even if this means some impositions on corporate and private behavior. Laws and regulations to carry out that role are regarded as a necessary and proper part of civic life, however inexact and burdensome they may be, and in some sense it can be said that there are even environmental rights of citizens to certain basic natural amenities. Global and national security is threatened by the continuation of ecological abuse and ignorance—including increases in world population, atmospheric alteration, and industrial toxicity—far more than by traditional national and ethnic enmities, and nations have a requirement in some way to address this new reality. Industrial society itself, in both its capitalist and state-socialist variations, both North and South, is a source of much of the eco-peril with its demands for growth, development, consumption, resource exploitation, and progress, as disasters from Bhopal to Chernobyl have made clear. And the stakes are known now to be of the very highest: either the continuing sustenance of the biosphere at virtually every point, on which the fate of the human species inescapably depends, or the destruction of this fragile realm, this earth.

Next—and crucial for understanding the future as well as the past—one must consider the shortcomings of environmentalism. Beyond the particular battles lost or crises continuing, there are three general problems, all endemic to the process of social change in America, no doubt, but not the less serious for being familiar.

The first might be called structural. The people in the important mainstream organizations are very largely white and very largely well-off, the more so as you move from membership to board of directors, and there very largely male as well. Whether charges of "racism" and "elitism" against them are quite fair—and those have been leveled with some frequency since the eighties—it is true that their concerns have tended to mirror those of the white suburban well-to-do constituencies and that the kinds of people who have been attracted to the staffs have tended to be college graduates, often professionals, and of the same general milieu as the people they deal with in legislatures and boardrooms. Both because American environmentalism came from a traditional conservationism little concerned with urban problems and because it seemed to ignore immediate social justice issues, very few nonwhites were much interested in it as a primary campaign or have signed on with environmental staffs. (*Audubon* reported in 1991 that Audubon itself had 35 nonwhites in a work force of 320, only 13 in professional positions, the National Wildlife Federation 19 of 283 people in professional jobs, and the Wilderness Society 4 among 80.) This in turn has tended to make these organizations slow to take on certain urban issues (incineration, lead poisoning) even when manifestly environmental—Lois Gibbs started her Citizens' Clearinghouse for Hazardous Wastes in part because she couldn't interest mainstream groups in the Love Canal scandal in 1978—and slow to demand attention to such rural issues as pesticide application and uranium mining that primarily affect people of color.

Along with this has been a practice in some parts of the movement of working with and accepting money from corporate America that seems to compromise its loftier ideals, whether or not this is compensated for by "pragmatic" working arrangements or increased research and lobbying budgets. Though corporate funding is said to be less than 5 percent of most of the organizational budgets, and nowhere more than a fifth, the sources can be somewhat unpleasant: General Electric and Waste Management as donors to Audubon and NWF, oil majors including Amoco, Exxon, and Mobil to Audubon, NWF, and World Wildlife, chemical giants like Dow, Du Pont, and Monsanto to those same three and others. The effects of such bedfellowism are no doubt mixed, but some have raised questions of co-optation and diversion—"We think it dilutes the message," Greenpeace says—and it is hard to imagine that such important donors do not have at least some political influence, however hotly denied, on their beneficiaries.

A second shortcoming can be called institutional, a function of success, and rather neatly exemplified by the Audubon Society in the purposeful new style it chose for itself in early 1991. Already at half a million members, it decided to aim for 1.2 million within five years ("To increase our effectiveness," said president Peter Berle), to broaden its interests from birds and bird habitat to more "people-oriented" issues like toxic wastes and population control, and to become one of the important actors in "bringing about change through the government process." All well and good, but the inevitable trouble with such growth is that, as a *Newsweek* analysis of Audubon put it, it muddies "whatever purity of purpose it once had and, paradoxically, whatever effectiveness it craves"; also, when the organizational identities are thus blurred, style and image become all-important "because when it comes to substance, the groups have become as indistinguishable as sparrows." Moreover, organizations of such size seem to solidify into hierarchy and centralization, becoming "so big, so top-heavy, that to keep the apparatus running they have in many ways become like the institutions they battle." Half a million members there might be, but they had almost no voice in the operations of the New York headquarters, and there was no necessary accountability to them other than occasional reports in the monthly magazine;

priorities and policies thus come to be set by the few professionals, with the danger that they are based "on their fund-raising campaigns," as one Washington insider said, rather than on important immediate issues.

Nor was Audubon distinctive in this: similar charges were leveled in this period against much of the mainstream movement, particularly the "Beltway biggies" in Washington. Sierra Club activists dissatisfied with their organization's compromises and collaboration even formed a dissident group, the Association of Sierra Club Members for Environmental Ethics, which picketed the national board of directors at their annual meeting in 1991. As one disgruntled member of the new group summed it up: "They've all become just like Queen Victoria—old, fat, and unimaginative."

The third deficiency is, for want of a better word, ideological. By and large both the mainstream majors and the grass-roots NIMBYs tend to see environmental problems as isolated aberrations within a functioning system, correctable by regulation and enforcement, and not as inevitable by-products of an economic system based on the imperative of growth and the exploitation of resources, and governments designed to protect it. This means that they tend to confine themselves to piecemeal reforms rather than structural changes and to isolate such problems and their solutions from what might be called a political context. In practical terms, it means concern for the siting of toxic-waste dumps rather than halting the production of toxics, the advocacy of neighborhood recycling centers rather than the vast reduction of packaging, the writing of pollution-control regulations without changing the bureaucracy that failed to carry out the old ones, the campaign to make cars energy-efficient instead of reducing their numbers by millions, to cite a few examples. It means that the great hopes expressed on Earth Day 1990, borne on the rhetoric of how serious the crises are, how urgent the solutions, how committed the homemakers, workers, politicians, and executives, simply cannot be realized within the context of existing norms and givens.

True it is that this is entirely in the American grain, in which special-interest lobbying and electoral lesser-evil-ism are the characteristic ways that hard and serious issues get handled; true, too, that this is entirely characteristic of the American public, which tends by and large not to understand events in a political context or see individual problems as evidence of a failure in the system. But it is also true that this inevitably condemns the environmental movement to a kind of perpetual tinkerism and finger-in-the-dike-ism, a never-ending record of defeats mixed with victories and of victories that are always provisional insofar as they do not alter the values of the prevailing system.

Of course, radical environmentalism, or at least the best of it, does have the various ideological perspectives that allow it to see the perils of reformism and the need for systemic and structural change, but it, too, is not without problems. In addition to the factionalism endemic to the faithful, it has yet to discover a way to influence the complacent core of the American public except momentarily or to avoid relegation to those fringes of political life where it can be ignored, co-opted, or suppressed. And obviously the more it raises questions about the underlying values of the American system behind the environmental crises, or the civilization on which it rests, the more it is certain to be resisted, in the short run if not the long.

Thus the shortcomings of a movement, even as it became entrenched in the American landscape as few such movements before.

To gauge the future of such a movement adequately it is finally necessary to suggest at least the considerable forces ranged against it: an indication of its success, no doubt, but also an indication of its difficulties.

The backlash against environmentalism was naturally initiated by the corporations that had most to lose by its successes. Initially the tactics were essentially duplicates of those of the environmental groups themselves—political lobbying and public relations—in what was estimated in 1990 to be about a $500-million-a-year operation, behind such deceptive and well-funded fronts as the National Wetlands Coalition (oil drillers and developers) and the U.S. Council on Energy Awareness (the nuclear power industry). In the mid-eighties the mass-mail fund-raising technique was added, with one such outfit, a direct-mail marketer in Bellevue, Washington, claiming to raise more than half a million dollars a year using the environmental movement as "the perfect bogeyman." Then, in November 1991, 125 business groups and fronts organized under a single flag as the Alliance for America, heavily funded by timber cutters, oil drillers, ranchers, and other anti-environmental corporations who aim, in the words of its chief ideologue, Ron Arnold, "to destroy environmentalists by taking their money and their members." "The tables have turned," says Mary Bernhard of the U.S. Chamber of Commerce, "and there's a renewed commitment now on the part of the business community to be more 'pro-active'" and to "turn things against" the environmental movement.

Where such legitimate efforts have proved inadequate, at least some corporations have turned to tougher tactics, encouraged by such prominent conservatives as former Interior Secretary Watt, who publicly declared in 1990 that "if the troubles from environmentalists cannot be solved in the jury box or at the ballot box, perhaps the cartridge box should be used." Environmental activists coast to coast have reported enough specific examples of violence targeted against them—offices trashed, cars smashed, homes entered, death threats, the home of a Greenpeace researcher burned in Arkansas, the office of a National Toxics Campaign worker burgled in Denver, two Earth

First! workers firebombed in California—to leave no doubt that some kind of concerted private crusade was being waged, and at high stakes. One of the most egregious examples surfaced in late 1991 when a congressional committee discovered that the corporate managers of the Trans-Alaska Pipeline paid hundreds of thousands of dollars for a nationwide hunt to find and silence critics of the Alaska oil industry—complete with eavesdropping, theft, surveillance, sting operations, and conspiracy—and harassed its own employees to cover up leaks about its environmental and safety errors. "It seems to me to be a very dangerous trend," said Representative George Miller, whose committee exposed the plot.

Corporations often seem to have had a helping hand in their operations from various governmental agencies, and according to investigator Chip Berlet in *The Humanist*, "tactics used to harass activists include obvious surveillance, intimidation, anonymous letters, phony leaflets, telephone threats, police overreaction and brutality, dubious arrests, and other threatening actions." The FBI has targeted environmentalists as legitimate prey, particularly those connected with Earth First!: in addition to the $2 million undercover operation against Dave Foreman and others in Arizona (which ended in a harsh plea bargain in 1991), it cooperated with local police in actions against EF!'s Redwood Summer in 1990 and has regularly visited and harassed EF! activists across the country. The EPA itself has also had a hand in curtailing critics: after a group of environmentalists demonstrated against an EPA-approved incinerator in California, the agency circulated videotapes of the action and forwarded a copy to police in a Phoenix suburb who used it to target the same protesters at an incinerator hearing there, arresting them forcibly and without provocation in the hearing room.

None of this backlash—"in full swing" as of 1992, according to one scholar—has so far derailed the environmental train, but it has undoubtedly taken a toll, particularly among the more radical groups and particularly at the local level. One activist in Fort Bragg, California, who has been vocal against the logging practices of Georgia Pacific, says that a company boycott of her day-care center forced it to close; an anti-toxic worker, harassed by thefts and mysterious power outages, says that with all this "something weird" going on, "we can't build a mass-based movement." Some Earth First!ers have decided to lie low after FBI visits and telephone death threats, and the mood at the EF! national rally in 1991 was said to have been decidedly "paranoid and prickly," according to one participant. There are no signs that environmental activism is waning, even among local groups, but according to one Washington group that monitors reports of environmental harassment, there is a new sense abroad that "the stakes are higher."

Ultimately it seems that the prospects of the U.S. environmental movement, after its initial decades of expansion and embeddedness, will depend on just how profound and sweeping a role it sees for itself in the face of the environmental paradox.

It can, on the one hand, continue to operate largely as a reformist citizens' lobby, pressured on the fringes by more radical groups but for the most part willing to work within the system and reap the victories, and rewards, therein. In this role it is likely to put increasing emphasis on scientific breakthroughs and technofixes—Amory Lovins's low-energy light bulbs, hydrogen fuels, photovoltaics, pollution-free coal, deep-ocean waste dumping—and press for a greater proportion of both corporate and government money to be given to environmental research and development. It is apt to make much more of the power of tax policies in changing damaging ecological practices—pollution taxes, for example, as advocated by the Worldwatch Institute, or property taxes tied to corporate environmental performance, as in Louisiana—or starkly new budget priorities to advance the environmentally benign and retard the environmentally destructive. And it is likely also to champion "green power" in the marketplace in new and more extensive ways, such as the Green Seal and Green Cross product-approval ratings, authoritative (perhaps government) "green-labeling" of foods and products, increased pressure on consumer-product industries (particularly for recycled and recyclable goods and packages), and corporate ethics pledges (like the Valdez Principles set up in September 1989) of voluntary environmental good behavior. Wilderness issues will almost certainly remain secondary, except for tropical rainforest protection and the occasional fight over one West Coast timber operation or another.

Or the movement can, on the other hand, deepen and darken its analyses and criticisms, following the lead of the more serious of the radical environmentalists, and try to work for structural changes in the system, with more rapidity or less as the needs arise. In this role it would likely keep up pressure on corporate and bureaucratic malefactors in the Earth First! and Rainforest Action Network style, and might well devote resources toward a strong Green Party that could make an electoral third-party challenge as serious as those in parts of Europe. But its primary emphasis would no doubt be extracorporative and extraparliamentary, aiming to change public opinion in as broad a manner as possible, from weighty tome to flagrant harangue: books and articles, rallies and campaigns, speeches and films, civil disobedience and mass protests. In the eyes of the Institute for Policy Study, a left-wing think tank in Washington, this sort of path would lead the environmental movement "to confront some of the largest issues of social organization—health, community, political and corporate accountability, jobs, and the future of the economy," and force it "to work toward a common analysis of the system it is attempting to change":

> The promise of environmentalism is that of a society which runs on a safe, sustainable, and democratic use of its

resources. The task for environmentalists now is to find or invent the means—economic, technical, and political—to transform this society into that one.

Not an easy task, made all the more difficult by the urgency of the time.

In the fall of 1990 the Public Broadcasting Corporation aired a six-part series called *Race to Save the Planet*, a perhaps belated but elaborate and surprisingly hard-hitting analysis of the ecological perils facing the earth and the extent of the changes that would have to be made to avert them. "Can we change the way we live," it asked as its central, its guiding, question, "in order to save our planet from destruction?" It is the seriousness and the passion with which the environmental movement takes that question that will for the most part determine what shape it takes in the years ahead. For it is the question that goes to the very heart of the American, indeed the industrial, system, its values, its assumptions, its configurations fashioned by five centuries of modern Western civilization.

If it is to be answered in the negative, whether by environmentalists and allies who argue that it is erroneous or alarmist or futile, then it is unlikely that any movement, no matter how powerful and well heeled and popular, will outrace the accumulative forces, from ozone depletion to global warming, toxic pollution to despoliation, resource depletion to overpopulation, that are leading the human species, and with it most other forms of life, to ecocide. If it is to be answered in the affirmative, whether by those environmentalists who believe it possible or only wish it were, then the race to destruction just may be slowed and halted, and the elements of an ecological society, modest and biocentric, attentive to nature's laws and living lightly on the planet as if it were the only one available, might eventually be fashioned.

*   *   *

In Maryland's Montgomery County, just north of the nation's capital, a new school building a few years ago was named in honor of Rachel Carson, who lived and worked in the county for many years. In 1990 it was discovered that the ground the school sits on was saturated, by builders following approved state guidelines, with a highly toxic insecticide, Dursban TC, used against termites. Dursban, which can remain toxic for ten years or more, is a deadly organophosphate of just the kind that Carson had written about in *Silent Spring*, a suspected cause of birth defects and cancers, to which children are especially susceptible if exposed by ingesting, touching, or inhaling. Though the foundation of the Rachel Carson Elementary School is built of masonry (and little liable to termite attack, one would have thought), the county school system will have to use a monitoring device for years to come to determine if Dursban is indeed seeping into the building that bears the great environmentalist's name.

"The current vogue for poisons," she wrote in *Silent Spring*, "has failed utterly to take into account" the fundamental fact that "we are dealing with life—with living populations and all their pressures and counterpressures, their surges and recessions. . . . These extraordinary capacities of life have been ignored by the practitioners of chemical control who have brought to their task . . . no humility before the vast forces with which they tamper." And in the final paragraph of her book, she added:

The concepts and practices of applied entomology for the most part date from [the] Stone Age of science. It is our alarming misfortune that so primitive a science has armed itself with the most modern and terrible weapons, and that in turning them against the insects it has also turned them against the earth.

So the tocsin rang more than thirty years ago, sending alarms —though unheeded by the Montgomery County school board and its builders, though unheeded indeed by a chemical industry that has subsequently produced some 30,000 chemicals of varying degrees of toxicity, though unheeded by most of the scientific establishment including the entomological—that nonetheless awoke a generation and spawned a movement that has without doubt created a real and lasting green revolution in the land. How that movement fares, how it proceeds, depends more than anything else on how deeply and purposively it reacts to the essential resonating message of that tocsin, a message that goes beyond the issues of chemicalization, even of pollution, important as they are, to the question of how this industrial society of the twentieth century will come to learn how to live with nature; or, as Carson put it, the question of "whether any civilization can wage a relentless war on life without destroying itself, and without losing the right to be called civilized."

ACKNOWLEDGMENTS

My deep thanks to W. H. Ferry, David Gurin,
Joy Harris, Sandy Irvine, David Levine,
Jon Naar, Marcus Raskin and the Institute for
Policy Studies, Lorna Salzman, George Sessions,
and Margaret Hays Young.

Bibliography

ADAMS, John H., et al., *An Environmental Agenda for the Future*, Island Press (Washington, D.C.), 1985.

BAILES, Kendall E., *Environmental History*, University Press of America (Denver), 1985.

BOOKCHIN, Murray, and Dave Foreman, *Defending the Earth*, South End/Learning Alliance, 1991.

BORRELLI, Peter, *Crossroads*, Island Press (Washington, D.C.), 1988.

BRAMWELL, Anna, *Ecology in the Twentieth Century*, Yale University Press, 1989.

BURCH, William R., et al., eds., *Social Behavior, Natural Resources and the Environment*, Harper, 1972.

CALDWELL, Lytton K., *Citizens and the Environment*, Indiana University Press, 1960.

CARSON, Rachel, *Silent Spring*, Houghton Mifflin, 1962.

CATTON, William R., *Overshoot*, Illinois University Press, 1980.

COHEN, Michael P., *The History of the Sierra Club, 1892–1970*, 1988.

COMMONER, Barry, *The Closing Circle*, Knopf, 1971.

DAY, David, *The Environmental Wars*, Ballantine, 1989.

DEVALL, Bill, and George Sessions, *Deep Ecology*, Gibbs M. Smith (Salt Lake City), 1985.

DIAMOND, Irene, and Gloria Feman Orenstein, *Reweaving the World: The Emergence of Ecofeminism*, Sierra Club, 1990.

DOWIE, Mark, *World Policy Journal*, Winter 1991–92.

DUNLAP, Riley E., and Angela G. Mertig, *American Environmentalism: The U.S. Environmental Movement, 1970–1990*, Taylor & Francis (Washington, D.C.), 1992.

EHRENFELD, David, *The Arrogance of Humanism*, Oxford, 1978.

EHRLICH, Paul, *The Population Bomb*, Sierra Club, 1968.

Environmental Ethics, 1985–92.

FOREMAN, Dave, *Confessions of an Eco-Warrior*, Harmony/Crown, 1991.

Fortune, February 12, 1990.

FOX, Stephen, *John Muir and His Legacy: The American Conservation Movement*, Little, Brown, 1981.

GRAHAM, Frank, *The Audubon Ark: A History of the National Audubon Society*, Knopf, 1990.

———, *Since Silent Spring*, Houghton Mifflin, 1970.

GOLDSMITH, Edward, ed., *A Blueprint for Survival*, Penguin, 1972.

Greenpeace, January–March, 1991.

HAYS, Samuel P., *Beauty, Health, and Permanence: Environmental Politics in the U.S., 1955–1985*, Cambridge University Press, 1987.

LANIER-GRAHAM, Susan D., *The Nature Directory*, Walker, 1991.

LEOPOLD, Aldo, *A Sand County Almanac*, Oxford University Press, 1949.

MANES, Christopher, *Green Rage: Radical Environmentalism*, Little, Brown, 1990.

MCCARRY, Charles, *Citizen Nader*, Signet, 1972.

MCCORMICK, John, *Reclaiming Paradise: The Global Environmental Movement*, Indiana University Press, 1989.

MCKIBBEN, Bill, *The End of Nature*, Random House, 1989.

MEADOWS, Donella, et al., *The Limits to Growth*, NAL, 1972.

———, *Beyond the Limits*, Chelsea Green (Post Mills, Vermont), 1992.

MILLS, Stephanie, *Whatever Happened to Ecology?*, Sierra Club, 1989.

NAAR, Jon, *Design for a Livable Planet*, Harper, 1990.

NASH, Roderick, *The Rights of Nature: A History of Environmental Ethics*, Wisconsin University Press, 1989.

———, *Wilderness and the American Mind*, Yale University Press, 1969, 1973.

NATAPOFF, Sasha, *Stormy Weather: The Promise of the U.S. Environmental Movement*, Institute for Policy Studies (Washington, D.C.), 1989.

National Geographic, December 1988.

NICHOLSON, Max, *The Environmental Movement*, Hodder & Stoughton, 1970.

PAEHLKE, Robert, *Environmentalism and the Future of Progressive Politics*, Yale University Press, 1989.

PETULLA, Joseph M., *American Environmentalism: Values, Tactics, Priorities*, Texas A & M University Press, 1980.

———, *American Environmental History*, Boyd & Fraser (San Francisco), 1977.

PHILIP, HRH Prince, *The Environmental Revolution*, Overlook, 1978.

PORTER, Gareth, and Janet Welsh Brown, *Global Environmental Politics*, Westview (Boulder), 1991.

RUSSELL, Dick, *Nation*, March 27, 1987.

——— and Owen de Long, *E*, November–December 1991.

SALE, Kirkpatrick, *Dwellers in the Land: The Bioregional Vision*, Sierra Club, 1985.

———, *'GBH* (Boston), October 1990.

———, *Mother Jones*, November 1986.

———, *Nation*, April 30, 1990.

SCHUMACHER, E. F., *Small Is Beautiful*, Harper, 1973.

SEREDICH, John, *Your Resource Guide to Environmental Organizations*, Smiling Dolphins (Irvine, California), 1991.

STONE, Peter, *Did We Save the Earth at Stockholm?*, Earth Island Ltd., 1973.

Time, January 9, 1989.

WARD, Barbara, and René Dubos, *Only One Earth*, Norton, 1972.

Wilderness, Spring 1989.

WORSTER, Donald, *Nature's Economy: The Roots of Ecology*, Sierra Club, 1977.

————, ed., *The Ends of the Earth: Perspectives on Modern Environmental History*, Cambridge University Press, 1988.

ZUCKERMAN, Seth, *Nation*, October 18, 1986.

INDEX